T0207187

Financial Data Analytics with R

Financial Data Analytics with R: Monte-Carlo Validation is a comprehensive exploration of statistical methodologies and their applications in finance. Readers are taken on a journey in each chapter through practical explanations and examples, enabling them to develop a solid foundation of these methods in R and their applications in finance.

This book serves as an indispensable resource for finance professionals, analysts, and enthusiasts seeking to harness the power of data-driven decision-making.

The book goes beyond just teaching statistical methods in R and incorporates a unique section of informative Monte-Carlo simulations. These Monte-Carlo simulations are uniquely designed to showcase the reader the potential consequences and misleading conclusions that can arise when fundamental model assumptions are violated. Through step-by-step tutorials and real-world cases, readers will learn how and why model assumptions are important to follow.

With a focus on practicality, *Financial Data Analytics with R: Monte-Carlo Validation* equips readers with the skills to construct and validate financial models using R. The Monte-Carlo simulation exercises provide a unique opportunity to understand the methods further, making this book an essential tool for anyone involved in financial analysis, investment strategy, or risk management. Whether you are a seasoned professional or a newcomer to the world of financial analytics, this book serves as a guiding light, empowering you to navigate the landscape of finance with precision and confidence.

Key Features:

- An extensive compilation of commonly used financial data analytics methods from fundamental to advanced levels
- Learn how to model and analyze financial data with step-by-step illustrations in R and ready-to-use publicly available data
- Includes Monte-Carlo simulations uniquely designed to showcase the reader the potential consequences and misleading conclusions that arise when fundamental model assumptions are violated
- Data and computer programs are available for readers to replicate and implement the models and methods themselves

Jenny K. Chen graduated with a Master's and Bachelor's degree in the Department of Statistics and Data Science at Cornell University. With expertise honed through academic pursuits and her current role as a quantitative product manager at Morgan Stanley, she is particularly interested in the applications of statistical modelling in finance and portfolio management. She was the youngest published author at the Joint Statistical Meetings in 2016 and has published several research papers in statistical modelling and data analytics.

Financial Data Analytics with R

Monte-Carlo Validation

Jenny K. Chen

CRC Press is an imprint of the
Taylor & Francis Group, an **informa** business
A CHAPMAN & HALL BOOK

Designed cover image: © Jenny Chen

First edition published 2025
by CRC Press
2385 NW Executive Center Drive, Suite 320, Boca Raton FL 33431

and by CRC Press
4 Park Square, Milton Park, Abingdon, Oxon, OX14 4RN

CRC Press is an imprint of Taylor & Francis Group, LLC

Library of Congress Cataloging-in-Publication Data

Names: Chen, Jenny K., author.
Title: Financial data analysis with R : Monte-Carlo validation / Jenny Chen.
Description: First edition. | Boca Raton, FL : CRC Press, 2024. | Includes bibliographical references and index.
Identifiers: LCCN 2023058122 (print) | LCCN 2023058123 (ebook) | ISBN 9781032745114 (hardback) | ISBN 9781032741499 (paperback) | ISBN 9781003469704 (ebook)
Subjects: LCSH: Finance--Mathematical models. | Finance--Econometric models. | R (Computer program language)
Classification: LCC HG106 .G466 2020 (print) | LCC HG106 (ebook) | DDC 332.0151955--dc23
LC record available at https://lccn.loc.gov/2023058122
LC ebook record available at https://lccn.loc.gov/2023058123

ISBN: 978-1-032-74511-4 (hbk)
ISBN: 978-1-032-74149-9 (pbk)
ISBN: 978-1-003-46970-4 (ebk)

DOI: 10.1201/9781003469704

Typeset in Latin Modern font
by KnowledgeWorks Global Ltd.

Publisher's note: This book has been prepared from camera-ready copy provided by the authors.

Contents

List of Figures

List of Tables

Preface

It is with great enthusiasm that I present this book on financial data analysis in *R*, a culmination of my academic journey at Cornell University and professional experiences at Morgan Stanley in the dynamic field of Wealth Management. This book has been shaped by the intersection of rigorous coursework and real-world applications, and it is my sincere hope that it serves as a bridge between statistical theory and practical applications of the financial industry.

My time at Cornell University, particularly within the Statistics and Operations Research Departments, laid the foundation for this endeavor. I was exposed to a comprehensive curriculum that combined theoretical concepts with hands-on applications. This includes courses in *Financial Data Analysis with R*, *Biostatistics*, *Categorical Data*, and *Stochastic Calculus for Financial Engineering*, and more. The intricate nature of financial data and the growing demand for proficient data analysis skills in finance prompted me to delve deeper into the realm of *R* programming for financial analytics.

Often I would consolidate my school notes from these classes and write them down. In doing homework problem sets and projects, I always wondered about the importance and relevancy of assumptions in statistical modeling. To explore this further, I would experiment with simulations and document them into *RMarkdown* for future reference, should I ever need to revisit my notes down the line. These notes of mine eventually became the prototypes of this book.

This book aims to serve as a comprehensive guide for students, researchers, and practitioners seeking to harness the power of *R* for analyzing financial data. The journey begins with a solid foundation in both financial theory and *R* programming, gradually progressing to more advanced topics such as time series analysis, risk and risk management. The practical examples provided throughout the book are rooted in real-world financial scenarios, offering readers a bridge between academic concepts and practical applications.

Post-graduation, I joined Morgan Stanley as an analyst in Wealth Management. Whenever I had the time, I would return to my notes and supplement them with additional content from my job or elsewhere. All the work and effort became the foundation for this book.

I would like to particularly emphasize the Monte-Carlo validation sections of this book, which have provided me a better understanding of the content and I hope that will be the case for all of you too!

Structure of the Book

This book is structured in ten chapters from classical statistical modeling all the way to time series modeling. As the opening of this book, *Chapter 1* serves as a fundamental exploration to understand *R* language. From the very basics such as from acquiring *R* software, installing it, upgrading *R* packages, to using *R* for data management and simulating data to showcase the renowned *Central Limit Theorem* in statistics.

With a basic understanding of *R* from Chapter 1, Chapter 2 to Chapter 4 are designed to analyze financial data for continuous data. *Chapter 2* is for financial data analysis with *linear regression models*, which is the fundamental statistical model in financial data analysis. *Linear regression modeling* is commonly used to understand and quantify relationships between variables in finance, which can aid in making informed decisions and predictions. In this chapter, we first give an overview of *linear regression* modeling and then illustrate a step-by-step approach to use *R* to analyze publicly available data (i.e., *wages*) with both *simple linear regression* and *multiple linear regression*. Concluding *Chapter 2*, we undertake a series of Monte-Carlo simulation studies designed to show the potential misleading conclusions that arise when the fundamental assumptions underlying *linear regression* are violated. Through these simulation studies, we aim to illuminate the risks associated with disregarding the underlying assumptions and encourage a more informed approach to financial analysis using *linear regression*.

Not all relationships in financial data are linear. In fact, there are more nonlinear relationships in real-life financial applications. *Chapter 3* serves to transform some of the nonlinear models commonly used in finance to their equivalent linear forms. The resulting linear models can then be modeled with linear regression described in *Chapter 2*. In *Chapter 3*, we select two of the most commonly-used nonlinear financial models, i.e., *the financial compounding model* and the *noncurrent asset depreciation model* as examples. We then transform these exponential nonlinear models to their corresponding linear forms so that linear regression can be applied to estimate their parameters. At the end of this chapter, a Monte-Carlo simulation study is designed to provide a comprehensive understanding of the potential benefits and limitations associated with transforming these nonlinear models into linear models.

While transformations like logarithms can handle many nonlinear relationships, certain financial models are inherently nonlinear due to their intricate dynamics. *Chapter 4* delves into the associated nonlinear statistical modeling. We focus on the *logistic growth model* with applications to model the adoption and saturation of financial products, services, or market trends and illustrate nonlinear regression with the *R* function *nls*.

Switching from continuous outcomes in financial data analysis to non-continuous (i.e., *categorical*) data, Chapter 5 and Chapter 6 are designed to illustrate commonly used *logistic regression* and *Poisson regression. Chapter 5* focuses on binary/binomial outcomes with *logistic regression modeling.* Binary outcomes are variables that can take only two values and are often used to model financial events with a *yes/no* or *success/failure* nature. The *logistic regression* is a common method for modeling binary/binomial outcomes, where the logistic function transforms a linear combination of predictors into probabilities. *Logistic regression modeling* is the *workhorse* in financial analysis, which is part of the *generalized linear model.* In this chapter, we briefly describe the theory of *logistic regression* and then illustrate the use of the *R* function *glm* to analyze a dataset called *Smarket* from *R* package *ISLR2.* In order to provide a deeper understanding of *logistic regression,* Monte-Carlo simulation-based validation is designed to clarify the *Misconception about Linear Regression* as well as the *Misconception about Logit-Transformed Linear Regression.*

Chapter 6 is then used for *counts data* analysis, which is another important and commonly seen data type. *Counts data* represents the number of events that occur within a fixed interval of time or space, such as the number of trades executed in the stock market, the number of defaults in insurance industries, or the number of market events within a specific time period. This type of data is different from the binomial data in Chapter 5, which is handled by *logistic regression. Count data* often follows a *Poisson* or *negative-binomial distribution,* and therefore, analysis would use methods like *Poisson regression* or *negative binomial regression.* In *Chapter 6,* we briefly describe the theoretical background of *Poisson regression* and the associated extension to model *underdispersion* and *overdispersion* using quasi-Poisson regression. Two datasets are used to illustrate the modeling of count data with the *R* function *glm.* To conclude this chapter, we design and and conduct a Monte-Carlo simulation study with the purpose of investigating the effects of *dispersion* on the conventional *Poisson regression.* In this investigation, we examine how *overdispersion* can lead to Type-I errors by producing an excessive number of false-positive instances, and how *underdispersion* can lead to Type-II errors, characterized by an excessive prevalence of false-negative instances.

The next three chapters from *Chapter 7* to *Chapter 9* are designed for time series data analysis, which are essential in financial data analysis. *Chapter 7* is an introduction to *Autoregressive Integrated Moving Average* modeling, which is characterized by three main components: 1) *Autoregressive (AR)* component represented by *order* p to capture the relationship between time series data and its past values, 2) *Integrated* component represented by *order* d to denote a differencing degree needed to make the data *stationary* (i.e., remove trends), and 3) *Moving Average (MA)* component represented by *order* q to model the dependency on past forecast errors. We continually utilize the *wages* data used in Chapter 2 to illustrate the application of *ARIMA regression* techniques in the analysis of time series data. To conclude the chapter, we design and implement a Monte-Carlo simulation study to highlight the different conclusions that can arise when using *linear regression* versus *ARIMA regression*. It underscores the importance of choosing the appropriate modeling technique when dealing with time-dependent data.

Financial markets often exhibit a striking tendency known as volatility clustering, where periods of high volatility are followed by more periods of high volatility, and the same holds true for low-volatility periods. *Chapter 7* is introduced to effectively model this time-varying volatility in the world of finance with more advanced *Autoregressive Conditional Heteroskedasticity (ARCH)* and *Generalized Autoregressive Conditional Heteroskedasticity (GARCH)* models. To introduce the *ARCH model* and *GARCH model* in this chapter, we will make use of compiled historical *Volatility Index* daily data from January 2nd, 1990 to June 27th, 2023 to illustrate *GARCH* modeling using *R* over the classical *ARIMA* modeling described in Chapter 7.

Transitioning from *ARIMA* and *GARCH* models to *cointegration*, *Chapter 9* introduces *cointegration* to analyze long-term relationships and interactions between multiple time series. Cointegration can essentially generate a *stationary* process from multiple time series that are *non-stationary*, as discussed in Chapter 7. For this purpose, we will use two datasets to illustrate how to use *R* packages for *cointegration analysis*.

Financial statistical modeling is a cornerstone of both risk management and wealth management and it equips financial professionals with the tools to make data-driven decisions and risk assent, as well as evidence-informed and optimized financial strategies. With all the financial modeling and statistical analysis from Chapters 1 to 9, Chapter 10 concludes this book to give a brief discussion and introduction of the critical role of statistical financial modeling in risk and wealth management.

The book is written in *RBookdown*. All the datasets and *R* programs used in the book can be obtained from the author by email jennykc99@gmail.com or downloaded from my *LinkedIn* site http://linkedin.com/in/jennykechen.

Acknowledgments

I extend my gratitude to the faculty members at Cornell University who have been instrumental in shaping my academic journey. Their guidance and mentorship have been invaluable, and I am fortunate to have been part of a learning environment that encourages exploration and critical thinking. In particular, I'd like to acknowledge the inspiring enthusiasm of Professor David Matteson, Professor Martin Wells, and Professor Joe Guinness. Having a professor who really contextualizes the content and teaches with passion really makes a difference. I'd also like to extend this gratitude to my TAs, Karen Grigorian and Yuchen Lu, who helped me through my courses.

I also gratefully acknowledge the professional support of Dr. Joseph Cappelleri, Executive Director at Pfizer, for supporting me in my university and statistics journey (go Big Red!). Finally, I'd like to thank Mr. David Grubbs and his team from Chapman & Hall/CRC, who have made the publication of this book a reality. To everyone involved in the making of this book, I would like to say **Thank You!**

This book is written using *rBookdown* package (Xie, 2019) and we thank the package creator, Dr. Yihui Xie, to create such a wonderful package for public use. We also thank the *R* creators and the community to create a wonderful open-source *R* computing environment so that this book can be possible. **Thank You All!**

Jenny K. Chen, MS & BS (Cornell University)

Product Manager at Morgan Stanley Wealth Management Group

New York City, NY, USA

About the Author

Jenny K. Chen graduated with a Master's and Bachelor's degree in the Department of Statistics and Data Science at Cornell University. These formative years laid the foundation for her passion for unraveling the intricacies of data and translating them into meaningful insights. At Cornell, she conducted academic research and led statistical projects as part of the Cornell Engineering Data Science project team. Prior to her university studies, Jenny was the youngest published author and presenter at the 2016 Joint Statistical Meetings held in Chicago.

Jenny is currently a Product Manager at Morgan Stanley in their Wealth Management division. Her role involves building quantitative investment tools for ultra-high-net-worth and institutional clients. These investment tools help clients make informed decisions about their wealth over time via investment and estate planning recommendations, all of which are backed by personalized analytics. Previously, Jenny has worked as a data scientist at Google, where she led a team of data scientists to develop several prediction algorithms for the 2019 NCAA March Madness Basketball Tournament. She has also published several research papers in statistical modeling and data analytics.

Outside of work, Jenny enjoys working on her food and travel blog, exploring New York City with friends, and reading. She is an avid traveler and a certified PADI Divemaster who is always finding ways to explore the world above and below. Jenny is also involved in Project by Project New York, a national charity organization of social entrepreneurs that serves Asian American nonprofits in need by raising public awareness and capital.

Connect with Jenny on LinkedIn to stay updated on her latest statistical explorations and insights.

1

Introduction to R

As the opening of this book, we start with the fundamental exploration of *R* in this chapter, catering to those who are unfamiliar with the language. Our aim is to provide a foundational understanding of *R*, starting from the very basics – from acquiring the *R* software, installing it, to upgrading *R* packages. Subsequently, we delve into illustrating how *R* can be effortlessly employed for data management and for simulating data to effectively showcase the renowned central limit theorem in statistics.

We conclude the chapter with a brief summary and some recommendations for further reading and references. The primary objective of this chapter is to acquaint readers with the realm of *R*. For those already versed in *R*, feel free to bypass this chapter and proceed directly to any of the subsequent chapters.

1.1 What is *R*?

The genesis of *R* can be traced back to its initial development by Ihaka and Gentleman (1996). Both creators were associated with the University of Auckland, located in New Zealand. Emerging during the mid-1990s, *R* has been developed out of the efforts of these individuals.

Over the years, *R* has achieved remarkable popularity, establishing itself as not just a programming language but also a versatile platform for statistical computing and data sciences. Its accessibility, coupled with a growing number of packages and libraries, has contributed to its widespread adoption across various fields, from academia to industry.

What sets *R* apart is its open-source nature, enabling a collaborative community to contribute to its expansion and enhancement. This collective effort is exemplified by the diverse group of *R* developers and researchers from different institutions around the world who form the core team behind the continuous evolution of *R*. This distributed development model has led to a dynamic and responsive environment, where innovations are shared and refined across geographical boundaries.

DOI: 10.1201/9781003469704-1

In essence, *R* stands as a testament to the power of collaborative creation and showcases how a humble beginning at the University of Auckland has evolved into a globally embraced resource for statistical analysis, data manipulation, and visualization.

To obtain an introduction to *R*, go to the official home page of the *R* project at http://www.R-project.org and click *What is R?*:

> R is a language and environment for statistical computing and graphics. It is a GNU project which is similar to the S language and environment which was developed at Bell Laboratories (formerly AT&T, now Lucent Technologies) by John Chambers and colleagues. R can be considered as a different implementation of S. There are some important differences, but much code written for S runs unaltered under R.
>
> R provides a wide variety of statistical (linear and nonlinear modeling, classical statistical tests, time-series analysis, classification, clustering, ...) and graphical techniques, and is highly extensible. The S language is often the vehicle of choice for research in statistical methodology, and R provides an open source route to participate in that activity.
>
> One of the R's strengths is the ease with which well-designed publication-quality plots can be produced, including mathematical symbols and formulae where needed. Great care has been taken over the defaults for the minor design choices in graphics, but the user retains full control.
>
> R is available as Free Software under the terms of the Free Software Foundation's GNU General Public License in source code form. It compiles and runs on a wide variety of UNIX platforms and similar systems (including FreeBSD and Linux), Windows and MacOS.

For certain users, the phrase *free* software might trigger a sense of caution, implying software that could potentially be challenging to navigate, of lower quality, or constructed using procedures that lack validation or verification. This perspective might arise due to a historical association between free software and potential drawbacks.

However, for others, *free* software carries an entirely different meaning, one rooted in the principles of *open-source* software. In this context, free software refers to software that not only grants users the freedom to utilize it without financial constraints but also empowers them with the freedom to delve into the software's inner system and modify it according to their specific needs. This second interpretation aligns with the concept of *open-source* software, where transparency, collaboration, and customization are key tenets.

The *R* system embodies this latter interpretation of *free* software. It is built upon the fundamental principle of openness and adaptability. Users are not only encouraged to leverage the software for their statistical and data-related endeavors, but they are also welcomed to explore its codebase, enhance its functionality, and tailor it to address a wide spectrum of analytical challenges.

We now proceed to the steps for installing and using *R*.

1.2 Steps on Installing *R* and Updating *R* Packages

Broadly speaking, the *R* system is composed of two main components. The first component is referred to as the *R base system*, encompassing the essential framework of the *R* language and its fundamental libraries. The second component comprises an array of user-contributed *packages*, each catering to distinct specialized applications.

Both the *base system* and the *packages* can be acquired via the Comprehensive R Archive Network (CRAN), accessible through the following web link: http://CRAN.r-project.org. Installation of *R* system is described in the following sections.

1.2.1 First Step: Install *R Base System*

The *base system* is readily accessible for diverse platforms including *Linux, MacOS X, and Windows.* In the context of this book, we focus on demonstrating the utilization of *R* specifically within the *Windows* environment. *Windows* users can obtain the latest *R* version through the following link: http://CRAN.r-project.org/bin/windows/base/release.htm (At the time of finishing this book, version *R4.3.3* will be available, which will be updated periodically).

Initiating the process to acquire and install *R* on your computer is straightforward. You can proceed by following the guidance provided in the installer. By default, this will lead to the installation of *R* into the *Program Files* subdirectory within your *C* drive. You can also specify the subdirectory you like to install *R*, which is what I usually do. Once completed, you are ready to enjoy the world of *R* for statistical computing and data analysis.

To illustrate *R* computation, suppose we wish to calculate the sum of 2017 (i.e., the year I started to write this book) and 7 (i.e., the years used to write and finish this book). The first line of *R* computation is:

```
# The Year started to write the book
yrStarted = 2017
# The years to write this book
yrWrite = 7
# The year finishing the book
yrFinish = yrStarted + yrWrite
```

The computed value, i.e., the year of this book published can be printed using:

```
yrFinish
```

```
## [1] 2024
```

You should get 2024, the year of the publication of this book! So you have done the first calculation in *R*!

1.2.2 Second Step: Installing and Updating *R* Packages

Within the foundation of the *R base system*, a large collection of standard statistical functions, descriptive and inferential statistical analysis techniques, as well as graphical capabilities are encompassed. These provisions are well-suited to address a wide spectrum of needs in the domain of statistical computing and data analysis.

You may install any *packages* from the *R* prompt by clicking *install.packages* from the *R* menu *Packages.*

Nevertheless, the *packages* constitute a realm of specialized applications that are created by proficient *R* users who possess expert knowledge in their respective domains. These *packages* are pivotal to the progression and enhancement of the *R* ecosystem. As of the writing of this book, the R system houses an impressive assemblage of over 21,000 packages, spanning virtually every facet of statistical computing and methodology. These invaluable resources are available for download through http://cran.r-project.org/web/packages/.

In fact, it's worth noting that within the expanse of *R*, the wealth of available packages is such that there is virtually no limit to what you can achieve. With this breadth of offerings, reassurance can be found in the assertion that *R* provides a comprehensive toolbox to meet virtually any analytical requirement.

For example, for financial analysts, there are many *R* packages for this purpose and there is a *CRAN Task View: Empirical Finance* maintained by *Dirk*

Eddelbuettel at https://CRAN.R-project.org/view=Finance. If you go to this webpage, you will see that this *CRAN Task View* included a list of packages useful for empirical work in Finance, grouped by topics in *Standard Regression Models, Time Series, Finance, Risk Management, Data and Date Management* with a list of *R* packages.

For example, for volatility modeling, the standard *GARCH(1,1) model* can be estimated with the *garch()* function in the *tseries* package. All the functionality of this package is then available by loading it to *R* as:

```
# Load the `tseries' package
library(tseries)
```

```
## Registered S3 method overwritten by 'quantmod':
##   method            from
##   as.zoo.data.frame zoo

##
##     'tseries' version: 0.10-54
##
##     'tseries' is a package for time series
##     analysis and computational finance.
##
##     See 'library(help="tseries")' for details.
```

For first-time users of this package, information about its use may be obtained by invoking the *help* manual, such as:

```
library(help=tseries)
```

Subsequently, an informative *help* page becomes accessible, elucidating the complete functionality encapsulated within the package. For those readers seeking an exhaustive compilation of the packages at their disposal, a visit to http://CRAN.R-project.org/src/contrib/PACKAGES is highly recommended.

1.2.3 Steps to Get Help and Documentation

A striking attribute of *R* lies in its seamless accessibility to the *Help and Documentation* resources, setting it apart from other software systems. The avenues for accessing *Help and Documentation* within *R* are diverse and varied. A general help for *R* can be obtained by typing *help.start{}* where you can find help on:

1. *Manuals* on
    - An Introduction to *R*
    - The R Language Definition
    - Writing *R* Extensions
    - *R* Installation and Administration
    - *R* Data Import/Export
    - *R* Internals
2. *Reference* on
    - Packages
    - Search Engine & Keywords
3. *Miscellaneous Material* about
    - *R*
    - Authors
    - Resources
    - License
    - Frequently Asked Questions
    - Thanks
    - NEWS
    - User Manuals
    - Technical papers
4. *Material specific to the Windows* port
    - CHANGES
    - Windows FAQ

For a comprehensive point of reference, *RGui* within *R* proves invaluable. Upon launching *R*, simply click on *Help* to gain access to an array of informative resources, including items like *FAQ on R on Windows*, and *Manuals (in PDF)*. In particular, we suggest that readers consider printing the online *PDF manual* titled *Introduction to R* for their own future reference.

Supplementary *Help and Documentation* resources can be accessed via the *R Homepage*. Extensive documentation and interactive discussions pertaining to *R* are readily accessible through the *R* homepage at http://www.r-project.org/. Notably, the online Documentation section encompasses an extensive collection of *manuals, FAQs, the R Journal, books, and other pertinent information*. It is advisable for readers to allocate time to peruse these online documents, as doing so will facilitate the process of becoming acquainted with R.

The most convenient way to access *Help* is from the *R* command prompt. You can always obtain specific help information from the *R* command prompt by using *help()*. For example, if you want help on fitting an *ARIMA* model to univariate time series data in the library *tseries*, type:

```
help(arima)
```

This will load an information page on *Fit an ARIMA model to a univariate time series* containing relevant information. This includes the description of the function, detailed usage for the function and some examples on how to use this function.

1.3 Database Management and Data Manipulations

1.3.1 Data Management with *RMySQL*

Within *R*, a large number of *packages* are designed for database management and data manipulation. Among these, *MySQL* stands out as one of the most widely employed databases, and it can be freely accessed via http://mysql.com, accommodating a range of platforms and boasting user-friendly configuration and operation.

In sync with the utility of *MySQL*, an *R* package named *RMYSQL* is available. This package, skillfully maintained by *Jeffrey Horner*, is accessible through http://cran.r-project.org/web/packages/RMySQL/. Especially for readers acquainted with *MySQL* and relational databases, we extend a strong recommendation for the adoption of this *R* package. Its capabilities encompass the creation of tables and the seamless storage of data into the *MySQL* framework.

1.3.2 Data Management with Microsoft Excel and *R* Package *gdata*

For many readers, *Microsoft Excel* probably stands as the most commonly used and user-friendly tool for data management. In light of this, we aim to introduce *R*'s capabilities for handling *Excel* data.

Multiple avenues exist to access *Excel* databooks. The *gdata* package serves as a versatile *R* package designed to extract data from *Excel* databooks. This package can be acquired from http://cran.r-project.org/web/packages/gdata/.

The function *read.xls* within the *gdata* package facilitates the import of *Excel* datasheets into the *R* environment. However, as this function is linked to a module in the *perl* programming language (http://perl.org), an initial installation of *perl* onto your computer is required. You can easily achieve this by visiting http://perl.org; *perl* is freely available. For instance, you could install *perl* at the path *c:/myprograms/perl64*. With *perl* properly installed, *read.xls* performs a two-step process. It initially translates the *Excel* spreadsheet into a comma-separated values (CSV) file. Subsequently, it invokes another *R* function, *read.csv*, to ingest the *.csv* file's contents.

We recommend readers familiarize themselves with this process, provided *perl* is already installed on their computer. It's important to note that specifying the file path on your computer where the *Excel* databook is stored is necessary for this procedure.

1.3.3 Data Management with Microsoft Excel and *R* Package *xlsx*

An alternative *R* package, *xlsx*, can be employed to import Excel files into the *R* environment. Unlike *gdata*, this package is *java-based* and reliant on *Java*. In order to utilize this package, a prerequisite step involves installing *Java* on your computer, tailored to your system. This can be readily accomplished by obtaining the free download from http://www.java.com.

Once *Java* is in place, the subsequent steps involve the installation of two *R* packages: *rJava* and *xlsx*. These packages facilitate the functionality of *xlsx* within the *R* environment.

The *xlsx* package offers two key functions for reading both *xls* and *xlsx Excel* files: *read.xlsx* and *read.xlsx2*. The latter function, *read.xlsx2*, is particularly advantageous for larger files as it offers improved processing speed when compared to the *read.xlsx* function.

1.3.4 Data Management with Microsoft Excel and *R* Package *readxl*

Distinguishing itself from the pre-existing packages such as *gdata* and *xlsx*, the *readxl* package, crafted by *Hadley Wickham* and his collaborators, boasts a unique feature. It requires no external dependencies on other packages, setting it apart as an easy-to-install and user-friendly solution applicable across all operating systems. This package is specifically designed for handling tabular data extracted from *Excel* files (i.e., *xls* or *xlsx* formats). Comprehensive information about its capabilities can be accessed at https://readxl.tidyverse .org.

1.3.5 Default Methods to Read Data into *R*

If you have *Microsoft Excel* installed on your computer, a simple approach to access the data is to first export it into a *tab-delimited* or *comma-separated* format. Subsequently, you can utilize *read.table* or *read.csv* functions within *R* to import the data seamlessly.

The *read.table* function stands as a versatile and adaptable method employed to import data into *R*, structuring it into a dataframe format.

To get familiar with its full functionality, use *help* as follows:

```
help(read.table)
```

You will see the description and detailed usage. Examples are given on how to use this function. Some of the functionalities are re-iterated here for easy reference:

- *header=TRUE*: If the first line in the data is the variable names, the *header=TRUE* argument is used in *read.table* to use these names to identify the columns of the output dataframe. Otherwise *read.table* would just name the variables using a *V* followed by the column number. In this case, we can use *col.names=* argument in *read.table* to specify a name vector for the dataframe.
- *row.names=*: The *row.names* argument is used to name the rows in the dataframe from *read.table*. Without *row.names* or with *row.names=NULL*, the rows in the dataframe will be listed as the observations numbers.
- *Missing Values*: As its default, *read.table* automatically treats the symbol *NA* to represent a missing value for any data type. For numeric data, *NaN*, *Inf* and *-Inf* will be treated as missing. If another structure is used for missing values, the *na.strings* argument should be used to refer to that structure to represent missing values.
- *skip=*: The *skip=* argument is used to control the number of lines to skip at the beginning of the file to be read; which is useful in situations where the data have embedded explanation at the beginning. For a very large datafile, we can specify *nrows=* argument to limit the maximum number of rows to read and increase the speed of data processing.

As wrappers for *read.table*, there are three functions of *read.csv*, *read.csv2*, and *read.delim* used specifically for comma-, semicolon-, or tab-delimited data, respectively.

1.3.6 *R* Package *foreign*

To accommodate various data formats beyond the standard ones, the *R* core team has developed a package named *foreign*. This package is designed for reading and writing data from diverse statistical software, including *Minitab*, *S*, *SAS*, *SPSS*, *Stata*, and *Systat*, among others. You can access this package through http://cran.r-project.org/web/packages/foreign/, and its comprehensive functionalities are elaborated in detail within the manual available at http://cran.r-project.org/web/packages/foreign/foreign.pdf.

1.4 Monte-Carlo Simulation Illustration

To exemplify the fundamental applications of *R* functionalities, we will design a Monte-Carlo simulation study. This study will make use of a diverse range of *R* features, encompassing data generation, data management, looping, graphics, and more. The primary objective is to showcase the well-known "Central Limit Theorem" in statistics through *R* capacities. This practical exercise will not only offer a better understanding of the fundamental central limit theorem in statistics but will also highlight the robust capabilities of *R* in facilitating both theoretical investigations and practical insights into real-world scenarios.

Let's start with the description of the *Central Limit Theorem.*

1.4.1 Central Limit Theorem

The *Central Limit Theorem (CLT)* is a fundamental concept in statistics. It states that the distribution of the sample means of a sufficiently large number of *independently and identically distributed* random variables approaches a *normal distribution*, regardless of the *original distribution* of the individual random variables. In simpler terms, the *CLT* explains why the *normal distribution* is often observed in real-world financial data, even if the underlying data may not be normally distributed.

Here are the key points of the *Central Limit Theorem*:

- **Random Sampling**: The CLT assumes that you have a random sample of observations from a population. Each observation should be independent of the others, and they should all be drawn from the same distribution, which we will show how to generate random sample using *R*.
- **Sample Size**: The sample size should be sufficiently large. There is no strict rule for what constitutes *sufficiently large*, but generally, a sample size of around 30 or more is often considered adequate.
- **Independence**: The observations in the sample should be independent of each other. This means that the outcome of one observation does not affect the outcome of another.
- **Identical Distribution**: Each observation in the sample should be drawn from the same distribution, regardless of what that distribution is.

The main consequence of the *Central Limit Theorem* is that as sample size increases, the distribution of the *sample means* becomes increasingly close to a *normal distribution*, regardless of the original distribution of the population. This *normal distribution* is centered around the true *population mean* and has a *standard deviation* that is related to the *population standard deviation* and the *sample size.*

The importance of the *Central Limit Theorem* lies in its application to *inferential statistics.* It allows us to make various assumptions and perform *statistical tests* and *confidence interval* calculations even when the population distribution is not known or not normal, as long as the sample size is sufficiently large. This principle underlies many statistical techniques used in hypothesis testing and confidence interval estimation.

More theoretically, let's assume that there is a set of independent and identically distributed (*i.i.d*) random variables $X_1, X_2, \cdots, X_n$ with a common mean μ and a common standard deviation σ. The *sum* of these random variables can be represented as:

$$S_n = X_1 + X_2 + \cdots + X_n.$$

Then the *average* or *sample mean* of these variables is given by:

$$\bar{X}_n = \frac{S_n}{n}.$$

According to the Central Limit Theorem:

- **The distribution of $\bar{X}_n$** approaches a normal distribution as n (the sample size) becomes larger, regardless of the distribution of the individual X_i variables.
- **The mean of the sample means,** $E(\bar{X}_n)$, is equal to the population mean μ.
- **The standard deviation of the sample means,** $SD(\bar{X}_n)$, is given by $\frac{\sigma}{\sqrt{n}}$, where σ is the population standard deviation and n is the sample size.

Mathematically, this can be expressed as:

$$\lim_{n \to \infty} P\left(\frac{\bar{x}_n - \mu}{\frac{\sigma}{\sqrt{n}}} \leq x\right) = \Phi(x),$$

where P represents the probability, $\Phi(x)$ is the cumulative distribution function of the standard normal distribution, and x is a specific value.

In practical terms, this central limit theorem is used to approximate the distribution of sample means. For instance, if you have a population with an unknown distribution, as long as your sample size is sufficiently large, you can use the normal distribution to make inferences about the sample mean. This is particularly useful for hypothesis testing, confidence interval estimation, and other statistical analyses.

1.4.2 *R* Capacity on Random Number Generation

R has a wide range of functions to handle probability distributions and data simulation from these probability distributions. For any distribution, *R* has a corresponding set of defined functions for its *density function* (denoted by d), *cumulative distribution function* (denoted by p), *quantile function* (denoted by q) and *random generation*(denoted by r). For example, for the commonly used normal distribution, its *density function, cumulative distribution function, quantile function* and *random generation* with mean equal to *mean* and standard deviation equal to *sd* can be generated using the following *R* functions:

```
# Density function using "d"
dnorm(x, mean = 0, sd = 1, log = FALSE)
# Cumulative distribution function using "p"
pnorm(q, mean = 0, sd = 1, lower.tail = TRUE, log.p = FALSE)
# Quantile function using "q"
qnorm(p, mean = 0, sd = 1, lower.tail = TRUE, log.p = FALSE)
# Random number generation using "r"
rnorm(n, mean = 0, sd = 1)
```

where x or q the is vector of quantile, p the is vector of probabilities, n the is number of observations, *mean* is the vector of means, *sd* is the vector of standard deviations.

The above specification can be found using the *Help* function as follows:

```
# help on rnorm
help(rnorm)
```

There are similar sets of *d, p, q, r* functions for *log-normal distribution, exponential distribution, Poisson distribution, Weibull distribution, Cauchy distribution, beta distribution, uniform distribution, binomial distribution, t distribution, F distribution, hypergeometric distribution, Beta distribution*, etc.

Also there is a *sample* function for sampling from a vector and a *replicate* function for repeating a computation.

1.4.3 *R* to Prove (Empirically) the Central Limit Theorem

There are many publications and books to prove and discuss the central limit theorem in detail. Interested readers can do some Google searching to find the appropriate books. We recommend *Probability and Statistics* by Morris H. DeGroot and Mark J. Schervish (DeGroot and Schervish, 2011). This textbook is commonly used in introductory statistics courses and covers the *Central Limit Theorem* along with its proof. It's a comprehensive resource for understanding various statistical concepts. We also recommend *Mathematical Statistics with Applications* by Wackerly, Mendenhall, and Scheaffer (Wackerly et al., 2008). This textbook is another widely used resource that covers the *Central Limit*

Theorem along with its proof. It provides a comprehensive introduction to mathematical statistics. Another book *All of Statistics: A Concise Course in Statistical Inference* by Larry Wasserman (Wasserman, 2003) covers a detailed discussion of the *Central Limit Theorem* and its proof. It's known for its clarity and concise presentation.

In this section, we illustrate the concept of the *central limit theorem* empirically using *R* to help the understanding of this theorem. There are numerous probability distributions to warrant exploration, each with its unique characteristics in financial data analysis. In this illustration, we focus on three frequently employed probability distributions within financial data analysis: the *normal distribution, log-normal distribution*, and *uniform distribution.* It's worth noting that interested readers with a penchant for other probability distributions can readily apply the same rationale and *R* code segments in this section to investigate distributions of their interest.

1.4.3.1 Normal Distribution

Normal distribution serves as the foundation in statistics and any financial modeling. In this section, we simulated a set of *independently and identically distributed* (*i.i.d*) normal random sample $(X_1, X_2, \cdots, X_n)$ with a common mean μ and a common standard deviation σ. Based on the central limit theorem, the *distribution of* $\bar{X}_n$ approaches a normal distribution with the mean μ and standard deviation of $\frac{\sigma}{\sqrt{n}}$.

To show the distribution of $\bar{X}_n$, we first simulate data from a standard normal distribution where $(\mu, \sigma) = (0, 1)$ with $n = 100$ using *rnorm* function and graphically show this distribution, which should look like a normal distribution.

We calculate the *mean* of these $n = 100$ samples and *replicate* this simulation many times (say *nsimu=10,000*) and we should have *nsimu=10,000* means. Based on the central limit theorem, the resulted sample distribution of these means should be normally distributed with mean equal to the population mean $\mu = 0$ and standard deviation equal to the $\frac{\sigma}{\sqrt{n}} = \frac{1}{\sqrt{100}} = \frac{1}{10}$

This can be shown from the following *R* code chunk:

```
# Set random seed for reproducibility
set.seed(333)
# Number of observations
n = 100
# Use "rnorm" to simulate one set of data
x0 = rnorm(n)
# Calculate its mean&sd:should be close to 0 and 1
mean(x0);sd(x0)
```

```
## [1] -0.02278
```

```
## [1] 0.9584
```

```
# Replicate "nsimu" times using "replicate" function
nsimu =10000
x= replicate(nsimu,mean(rnorm(n)))
# Check the number of replication: should be 10000
length(x)
```

```
## [1] 10000
```

It is evident that the generated random normal sample *x0* exhibits a mean of -0.02278394 and a standard deviation of 0.9584083. These values closely approximate the parameters of the standard normal distribution, where the mean is 0 and the standard deviation is 1. To further illustrate the concept of the central limit theorem, we employ the *replicate* function in *R* to iteratively "replicate" the mean *mean(rnorm(n))* for *nsimu=10,000* iterations. We assign the resulting values to the variable x, yielding a dataset with a length of 10,000. This signifies that we have generated 10,000 *sample means* from the standard normal distribution, serving as a representation for showcasing the central limit theorem.

We can now plot *x0* and x using the following *R* code chunk:

```
# Create the plot layout for two figures
layout(matrix(1:2,2,1))
# Plot the first sample
hist(x0, xlab="Histogram of Sample ('x0')
     from Standard Normal Distribution",
     main="Normal Distribution")
# Plot the means of the 10000 replications
hist(x, xlab="Histogram of the 10,000
     Sample Means ('x')",
     main="", freq=FALSE)
# Overlay the normal distribution based on CLT
curve(dnorm(x,0,1/sqrt(n)),col=2,
      lty=2,lwd=2,add=TRUE)
```

In Figure 1.1, the top panel is the histogram of the sample with sample size of $n = 100$ drawn from the standard normal distribution. It shapes like the standard normal distribution even it is not perfect due to the sample size of 100. The bottom panel is the histogram of the 10,000 sample means calculated from the 10,000 draws from the standard normal distribution using the *R* code *x= replicate(nsimu,mean(rnorm(n)))* and we overlaid the normal distribution with mean 0 and standard deviation of *1/sqrt(n)*. It can be seen that the

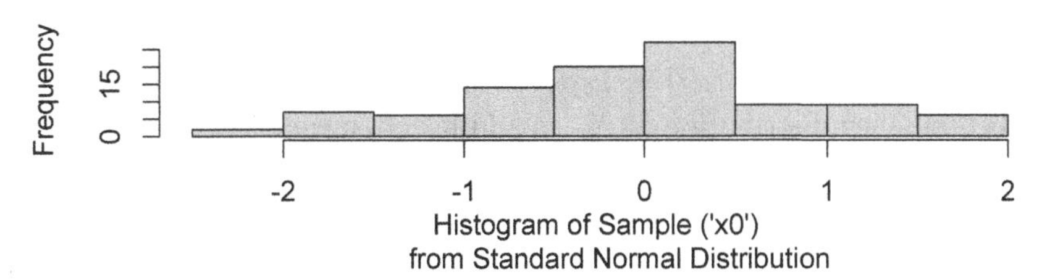

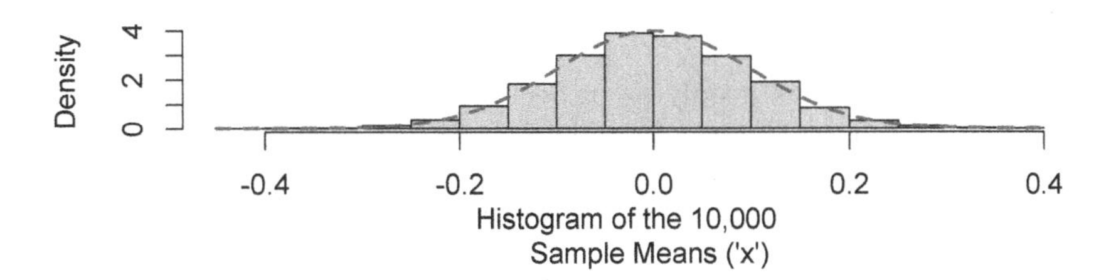

FIGURE 1.1
Illustration of Data Generated from Standard Normal Distribution

histogram distribution behaves like the normal distribution with mean 0 and standard deviation $\frac{1}{\sqrt{n}} = \frac{1}{\sqrt{100}} = \frac{1}{10}$. This confirms the central limit theorem.

1.4.3.2 Log-Normal Distribution

The log-normal distribution is another significant probability distribution commonly used in financial modeling to represent the distribution of asset prices, returns, and other financial variables. It is particularly useful for modeling variables that are subject to multiplicative growth or compounding over time, which is often the case in financial markets and will be discussed more in Chapter 3.

In a log-normal distribution, the logarithm of the variable follows a normal distribution. Mathematically, if X is log-normally distributed, then $ln(X)$ (natural logarithm of X) follows a normal distribution. The log-normal distribution is also characterized by two parameters: the mean (μ) and the standard deviation (σ) of the logarithm of the variable X. Specifically, if $ln(X)$ is normally distributed with the mean (μ) and the standard deviation (σ), then X is log-normally distributed. In this distribution, the associated mean of X is $E(X) = \exp\left(\mu + \frac{\sigma^2}{2}\right)$ and the variance is $Var(X) = \left[\exp\left(\sigma^2\right) - 1\right] \times exp\left(2\mu + \sigma^2\right)$.

In this demonstration, we will simulate data from the standard log-normal distribution with $\mu = 0$ and $\sigma = 1$, i.e., $ln(X) \sim N(0,1)$. Therefore X is log-normal with mean $E(X) = \exp\left(\mu + \frac{\sigma^2}{2}\right) = \exp\left(\frac{1}{2}\right)$ and $Var(X) = \left[\exp\left(\sigma^2\right) - 1\right] \times exp\left(2\mu + \sigma^2\right) = \left[\exp\left(1\right) - 1\right] \times exp\left(1\right)$. Based on the central limit theorem, the distribution of $\bar{X}_n$ should have a mean:

$$E(\bar{X}_n) = E(X) = \exp\left(\mu + \frac{\sigma^2}{2}\right) = \exp\left(\frac{1}{2}\right)$$

and

$$Var(\bar{X}) = \frac{Var(X)}{n} = \frac{\left[\exp\left(\sigma^2\right) - 1\right] \times exp\left(2\mu + \sigma^2\right)}{n} = \frac{\left[\exp\left(1\right) - 1\right] \times exp\left(1\right)}{n}.$$

While we won't delve as deeply as in Section 1.4.3.1 which focuses on the normal distribution, we'll now consolidate these *R* code snippets for the log-normal distribution. You can find the integration of these code segments in the subsequent *R* code chunk, accompanied by thorough explanations:

```
# Set the random seed
set.seed(3388)
# Generate x0 from log-normal with mu=0, sigma=1
x0 = rlnorm(n)
# Generate 10,000 means from log-normal
x= replicate(10000,mean(rlnorm(n)))
# Create the plot layout for two figures
layout(matrix(1:2,2,1))
# Plot the first sample
hist(x0, nclass=20, xlab="Histogram of Sample ('x0')
     from Standard log-Normal Distribution",
     main="Log-Normal Distribution")
# Overlay the standard normal distribution
# curve(dlnorm(x0,0,1),col=2,lty=2,lwd=2,add=TRUE)
# Plot the means of the 10000 replications
hist(x, xlab="Histogram of the 10,000 Sample
     Means ('x')", main="", freq=FALSE)
# Overlay the normal distribution based on CLT
curve(dnorm(x,exp(1/2),
            sqrt((exp(1)-1)*exp(1))/sqrt(n)),
      col = 2, lty = 2, lwd = 2, add = TRUE)
```

As seen in Figure 1.2, the top panel is the histogram of the sample with sample size of $n = 100$ drawn from the standard log-normal distribution, which is a skewed distribution. The bottom panel is the histogram of the 10,000

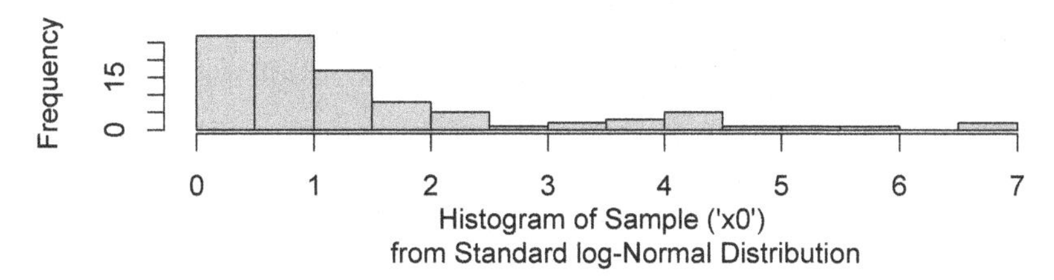

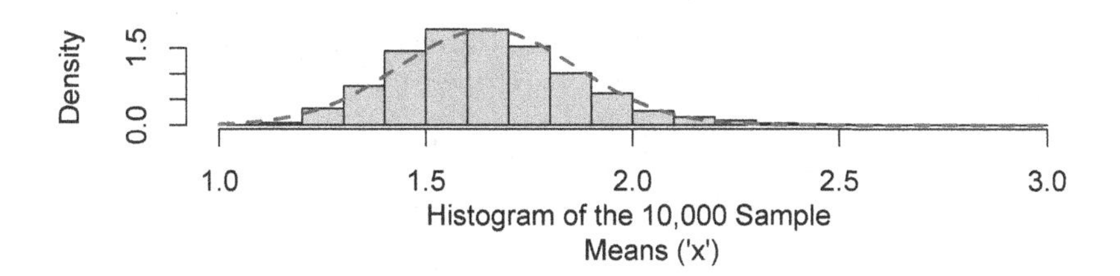

FIGURE 1.2
Illustration of Data Generated from Log-Normal Distribution

sample means calculated from the 10,000 draws from the standard log-normal distribution using the *R* code *x= replicate(nsimu,mean(rlnorm(n)))* and we overlaid the normal distribution with mean $E(\bar{X}_n) = \exp\left(\frac{1}{2}\right)$ and standard deviation of $\sigma_{\bar{X}_n} = \sqrt{Var(\bar{X}_n)} = \sqrt{\frac{[\exp(1)-1]\times exp(1)}{n}}$. It can be seen that the histogram distribution behaves like the normal distribution and this again confirms the central limit theorem for log-normally distributed data.

1.4.3.3 Uniform Distribution

The *uniform distribution* is another important probability distribution that has applications in financial modeling, particularly when dealing with scenarios involving randomness, fairness, or equally likely outcomes. The *uniform distribution* has been used in certain *option pricing models* to simulate the random movement of underlying assets or factors that influence option values. In *risk management*, the *uniform distribution* was used in *stress testing* or *sensitivity analysis* to examine how financial models react to extreme or unexpected events. In *capital budgeting decisions*, the *uniform distribution* can model uncertain cash flows or rates of return for different projects.

In a *uniform distribution*, all outcomes in a given range are equally likely to occur. It is characterized by two parameters: the minimum (a) and maximum (b)

values of the range. The probability density function of the uniform distribution ensures that each value in the range has the same probability of being chosen.

Mathematically, if X is uniformly distributed within interval $[a, b]$, then $X \sim U[a, b]$ with probability density function (pdf) defined as follows:

$$f(x) = \begin{cases} \frac{1}{b-a} & x \in [a, b] \\ 0 & \text{otherwise} \end{cases}$$

With this uniform distribution, the mean is $E(X) = \frac{a+b}{2}$ and the associated variance is $Var(X) = \frac{(b-a)^2}{12}$. If we take samples from this *uniform distribution* and calculate the *mean*, the distribution of these *means*, will be normally distributed with mean as $E(\bar{X}_n) = E(X) = \frac{a+b}{2}$ and the associated variance is $Var(\bar{X}_n) = \frac{Var(X)}{n} = \frac{(b-a)^2}{12n}$ based on the central limit theorem, i.e., $\bar{X}_n \sim N\left(\frac{a+b}{2}, \frac{(b-a)^2}{12n}\right)$.

We will show this property using *R* Monte-Carlo simulation as what we did before for *normal distribution* and *log-normal distribution*. For this illustration, we choose $a = -1$ and $b = 1$, i.e., $X \sim U[-1, 1]$. Then $E(X) = \frac{a+b}{2} = \frac{0}{2} = 0$ and the associated variance is $Var(X) = \frac{(b-a)^2}{12} = \frac{1}{3}$. Further $E(\bar{X}_n) = E(X) = 0$ and the associated variance is $Var(\bar{X}_n) = \frac{Var(X)}{n} = \frac{(b-a)^2}{12n} = \frac{1}{3n}$. The following *R* code chunk is used for this purpose with detailed explanation which will produce Figure 1.3 to graphically show the *central limit theorem* for *uniform distribution*.

```
# The plot layout
layout(matrix(1:2,2,1))
# Set random seed
set.seed(333)
# Generate one sample from uniform distribution
x0 = runif(n,-1,1)
# Plot the uniform distribution
hist(x0, xlab="U(-1,1)", main="Uniform Distribution")
# Replicate the mean calculation for 10,000
nsimu = 10000
x = replicate(nsimu,mean(runif(n,-1,1)))
# Plot the mean distribution from 10,000 simulations
hist(x, xlab="Distribution of the Means from U[-1,1]",
     main="", freq=FALSE)
# Overlay the normal distribution
curve(dnorm(x,0,sqrt(1/3)/sqrt(n)),
       col = 2, lty = 2, lwd = 2, add = TRUE)
```

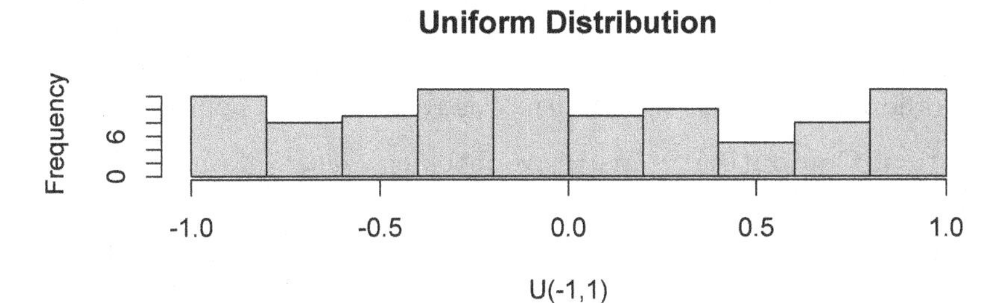

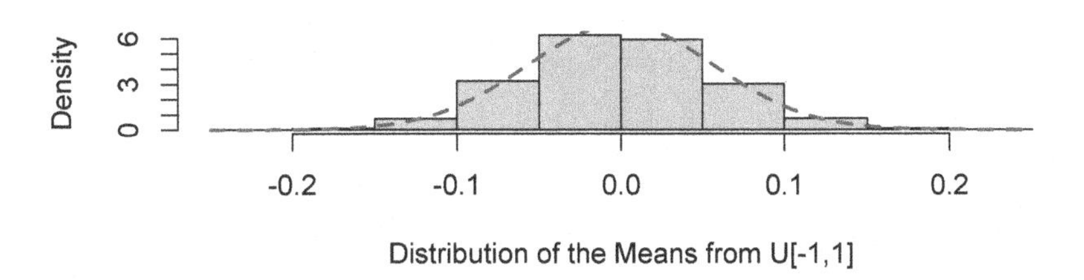

FIGURE 1.3
Illustration of Data Generated from Uniform Distribution

1.5 Summary and Recommendations for Further Reading

In this chapter, we've introduced the world of the *R* system, covering its installation process and the associated packages. We've showcased how *R* can be utilized for data simulation, manipulation, and statistical graphics through the Monte-Carlo simulation to the well-known central limit theorem in statistics.

For those who wish to delve deeper into the *R* system and enhance their familiarity, we offer the following recommendations:

- **R Fundamentals and Programming**: To grasp the intricacies of the *R* language and its programming structures, the books by *John Chambers* (Chambers, 1998) and (Chambers, 2008) provide exceptional insights.
- **R Graphics**: For a comprehensive understanding of *R* graphics, not only do we endorse Sarkar's book (Sarkar, 2008) on *lattice*, but we also recommend delving into *Paul Murrell*'s book (Murrell, 2005) as well as the wonderful *ggplot* by Hadley Wickham (Wickham, 2016).

- **Statistical Data Analysis with R**: To harness *R* for statistical modeling, Faraway's books from 2004 (Faraway, 2004) and 2006 (Faraway, 2006) are invaluable resources. Additionally, the book by *Everitt and Hothorn* (Everitt and Hothorn, 2006) offers an excellent classroom-oriented approach.
- **Statistical Computing**: Maria Rizzo's book on *Statistical Computing with R* (Rizzo, 2008) provides profound insights into this domain.
- **User-Friendly Learning**: For those seeking a more accessible introduction to *R*, books such as (Kabacoff, 2011), (Gardener, 2012), and (Adler, 2012) offer user-friendly tutorials, practical problem-solving, and comprehensive references.
- **R Online Documentation**: Emphasizing the wealth of resources available, we reiterate the importance of exploring the numerous free online books, manuals, journals, and more accessible via the R homepage under the Documentation section.

By delving into these resources, readers can significantly enhance their understanding and proficiency with the *R* system, equipping themselves with the knowledge to wield its capabilities effectively.

1.6 Exercises

1. Investigate the *Central Limit Theorem* for t-distribution with small degrees of freedom of $df = 5$, medium degrees of freedom of $df = 50$, and large degrees of freedom of $df = 500$.
2. Investigate the *Central Limit Theorem* for Cauchy distribution (note that the *Cauchy distribution* has no finite moments).

2

Linear Regression

With the basic understanding of *R* from Chapter 1, we now proceed to financial data analysis with statistical methods. Regression modeling is the fundamental statistical model in financial data analysis and it plays a crucial role in understanding and quantifying relationships between variables in finance, which can aid in making informed decisions and predictions.

In this chapter, we will give an overview of *linear regression* modeling. We illustrate the regression technique using a public available financial data *wages*, which was originally studied by Nicols (1983), and later used in Krämer and Sonnberger (1986). This dataset is available in *R* library *lmtest* to illustrate regression diagnostics. A step-by-step illustration will be implemented with *R* to analyze this *wages* data with both simple linear regression (*SLR*) and multiple linear regression (*MLR*).

Concluding this chapter, we will undertake a series of simulation studies designed to show the potential misleading conclusions that arise when the fundamental assumptions underlying this *linear regression* are violated. The objective of these simulations is to understand the significance of avoiding misinterpretations in financial data analysis when utilizing *linear regression* techniques. Through these cautionary simulation exercises, we aim to illuminate the risks associated with disregarding the underlying assumptions and encourage a more informed approach to financial analysis using *linear regression.*

2.1 Descriptive Data Analysis

In this data, a yearly time series from 1960 to 1979 with four variables as follows:

- *w*: wages,
- *CPI*: consumer price index,
- *u*: unemployment,
- *mw*: minimum wage.

DOI: 10.1201/9781003469704-2

The data is available in *R* package *lmtest*, which can be loaded into *R* as follows:

```
# Load the library
library(lmtest)


# Load the data into R
data(wages)
# Check the dimension of the data
dim(wages)

## [1] 20  4

# Print data
wages

## Time Series:
## Start = 1960
## End = 1979
## Frequency = 1
##          w    CPI     u     mw
## 1960  NA  0.347    NA     NA
## 1961  NA  1.499    NA     NA
## 1962 2.8  1.477 6.457 15.000
## 1963 2.7  1.120 6.192  0.000
## 1964 2.7  1.107 5.763  8.696
## 1965 2.2  1.424 5.018  0.000
## 1966 2.9  1.188 4.170  0.000
## 1967 4.5  2.775 3.725 12.000
## 1968 5.1  2.700 3.723 14.286
## 1969 5.5  3.943 3.686  0.000
## 1970 6.2  5.058 3.556  0.000
## 1971 6.2  6.019 5.395  0.000
## 1972 6.3  4.629 6.272  0.000
## 1973 5.5  3.506 5.433  0.000
## 1974 6.2  4.677 4.522  0.000
## 1975 9.1 10.247 6.163 31.250
## 1976 7.6 10.273 8.625  9.524
## 1977 6.9  6.147 7.877  0.000
## 1978 7.5  6.388 6.679 15.217
## 1979 7.2  6.510 5.494  9.434
```

As seen from the output, this dataset contains 4 variables with time series of 20 observations from 1960 to 1979. Note that data for the years 1960 and 1961 are missing for *w*, *CPI*, *u*, *mw*, which are denoted by *NA*s. For illustration purposes, we will manipulate the original data *wages* to create a lagged series and a new dataframe *nwages* for this chapter using the following *R* code chunk:

```
# Transform to include lag-1 and lag-2 for CPI
nwages = cbind(wages, lag(wages[,2], k = -1),
               lag(wages[,2], k = -2))
# Name CPI1 = lag-1 and CPI2 = lag-2
colnames(nwages) = c(colnames(wages), "CPI1", "CPI2")
# Create a new dataframe for data from 1962 to 1979
nwages = window(nwages, start=1962, end=1979)
# Print the data
nwages
```

```
## Time Series:
## Start = 1962
## End = 1979
## Frequency = 1
##        w    CPI     u     mw   CPI1   CPI2
## 1962 2.8  1.477 6.457 15.000  1.499  0.347
## 1963 2.7  1.120 6.192  0.000  1.477  1.499
## 1964 2.7  1.107 5.763  8.696  1.120  1.477
## 1965 2.2  1.424 5.018  0.000  1.107  1.120
## 1966 2.9  1.188 4.170  0.000  1.424  1.107
## 1967 4.5  2.775 3.725 12.000  1.188  1.424
## 1968 5.1  2.700 3.723 14.286  2.775  1.188
## 1969 5.5  3.943 3.686  0.000  2.700  2.775
## 1970 6.2  5.058 3.556  0.000  3.943  2.700
## 1971 6.2  6.019 5.395  0.000  5.058  3.943
## 1972 6.3  4.629 6.272  0.000  6.019  5.058
## 1973 5.5  3.506 5.433  0.000  4.629  6.019
## 1974 6.2  4.677 4.522  0.000  3.506  4.629
## 1975 9.1 10.247 6.163 31.250  4.677  3.506
## 1976 7.6 10.273 8.625  9.524 10.247  4.677
## 1977 6.9  6.147 7.877  0.000 10.273 10.247
## 1978 7.5  6.388 6.679 15.217  6.147 10.273
## 1979 7.2  6.510 5.494  9.434  6.388  6.147
```

It can be seen from this output, there are two variables *CPI1* and *CPI2* in this new dataframe *nwages* where *CPI1* is the lagged-1 and *CPI2* is the lagged-2 from *CPI* in the original dataframe *wages*. Lag is essentially "delay". Just as correlation shows how much two time series are similar, autocorrelation describes how similar the time series is with itself. The relationship among these variables can be seen in Figure 2.1. From this figure, we can clearly notice that there is a strong relationship between wages (i.e., *w*) and consumer

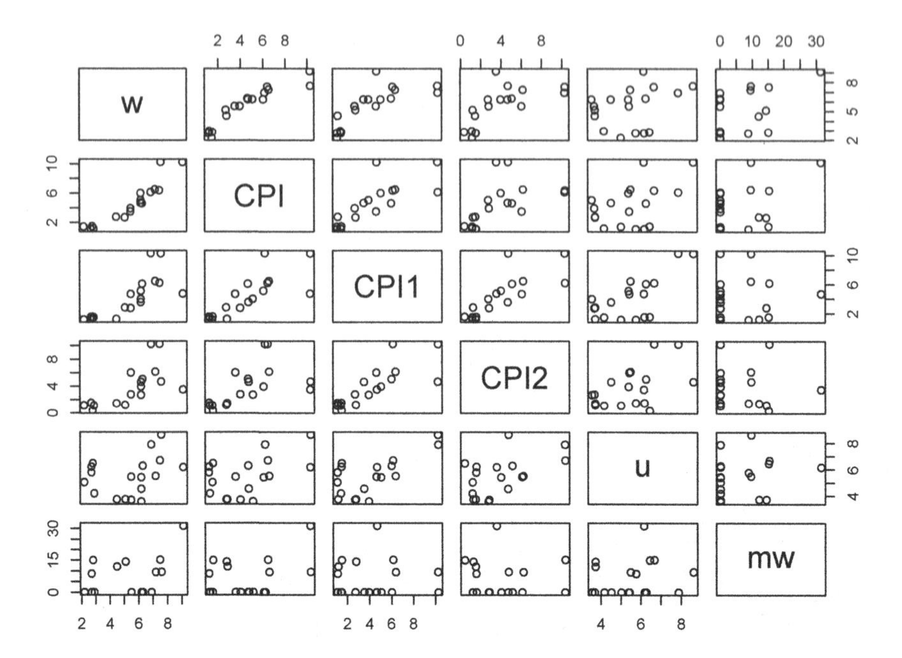

FIGURE 2.1
Relationships among Variables

price index (i.e., *CPI*, *CPI1*, *CPI2*), but less significant to unemployment (i.e., u) and minimum wage (i.e., mw). A formal statistical analysis is needed to firmly and statistically conclude this relationship.

2.2 Review for *Multiple Linear Regression* Models

Before we analyze this data, let us review (without formal deep theoretical proof) the basic *linear regression* theory in this section.

2.2.1 Multiple Linear Regression Model

Suppose that we have a data with n (i.e., $n = 20$ in *wages* data) observations in a study with outcome variable y which is related and predicted from K independent variables of $x_1, x_2, \cdots, x_K$. The general multiple linear regression (*MLR*) model can be written in the following equation:

$$y_i = \beta_0 + \beta_1 x_{1i} + \beta_2 x_{2i} + \cdots + \beta_K x_{Ki} + \epsilon_i \tag{2.1}$$

where $i = 1, \cdots, n$ for the ith observation.

For simplicity, we can put the MLR model in equation (2.1) into matrix,

$$y = X\beta + \epsilon \tag{2.2}$$

where y is a $n \times 1$ vector of the observed response variable, X is a $n \times (K+1)$ design matrix including all the independent variables with the first column as one for the intercept (i.e., β_0), $\beta = (\beta_0, \beta_1, \beta_2, \cdots, \beta_K)$ is a $(K+1)$-vector of regression parameters, and ϵ is the error term. Specifically,

$$y = \begin{pmatrix} y_1 \\ y_2 \\ \vdots \\ y_n \end{pmatrix}, \quad \epsilon = \begin{pmatrix} \epsilon_1 \\ \epsilon_2 \\ \vdots \\ \epsilon_n \end{pmatrix},$$

$$X = \begin{pmatrix} 1 & x_{11} & x_{21} & \cdots & x_{K1} \\ 1 & x_{12} & x_{21} & \cdots & x_{K2} \\ \vdots & \vdots & \vdots & \ddots & \vdots \\ 1 & x_{1n} & x_{2n} & \cdots & x_{Kn} \end{pmatrix}.$$

When there is only one variable in X (i.e., $K = 1$), the general *multiple linear regression* model in equation (2.1) is simplified as follows:

$$y_i = \beta_0 + \beta_1 x_{1i} + \epsilon_i \tag{2.3}$$

This equation is a simple linear regression (SLM), which is distinguished from the equation (2.1) with more than one variables in multiple linear regression.

Corresponding to the SLM in equation (2.3), the matrix form becomes as follows:

$$y = X\beta + \epsilon \tag{2.4}$$

where y is still a $n \times 1$ vector of the observed response variable, X is now a $n \times 2$ design matrix, and $\beta = (\beta_0, \beta_1)$. Specifically,

$$y = \begin{pmatrix} y_1 \\ y_2 \\ \vdots \\ y_n \end{pmatrix}, \quad X = \begin{pmatrix} 1 & x_{11} \\ 1 & x_{12} \\ \vdots & \vdots \\ 1 & x_{1n} \end{pmatrix}, \quad \epsilon = \begin{pmatrix} \epsilon_1 \\ \epsilon_2 \\ \vdots \\ \epsilon_n \end{pmatrix}.$$

2.2.2 The Method of Least Squares Estimation (LSE)

The *least squares estimation*(*LSE*) method is a crucial component of linear regression. *LSE* aims to minimize the sum of squared residuals (differences between observed and predicted values). This optimization strategy ensures that the chosen regression line best fits the data points.

Fundamentally, by estimating the parameter β, we are trying to find values (estimates) such that the systematic component (i.e., X) explains as much of the response variation (i.e., y) as possible. Hence, we are finding parameter estimates that make the error as *small* as possible. This is called *least squares estimation*; i.e., find β so that the error sum of squares is as small as possible. Therefore, the least squares estimate (LSE) of β, denoted by $\hat{\beta}$, can be obtained by minimizing the sum of squared errors (SSE):

$$\begin{aligned} SSE &= \sum_i \epsilon_i^2 = \epsilon'\epsilon \\ &= (y - X\beta)'(y - X\beta) \\ &= y'y - 2\beta X'y + \beta' X' X\beta. \end{aligned} \tag{2.5}$$

Taking the derivative of the sum of squared errors (SSE) with respect to β and setting to zero leads to

$$X'X\hat{\beta} = X'y. \tag{2.6}$$

When $X'X$ is *invertible*, we can obtain the LSE of β as follows:

$$\hat{\beta} = (X'X)^{-1}X'y. \tag{2.7}$$

In this LSE, the parameter estimation $\hat{\beta}$ can be estimated by equation (2.7). This estimation is simply calculated with the above equation and is therefore mathematically guaranteed globally with the desired solution. By *Gauss-Markov Theorem*, the LSE is the best linear unbiased estimator ($BLUE$) under the *Gauss-Markov theorem*. This theorem establishes that, within the class of linear estimators, LSE has the smallest variance among unbiased estimators.

2.2.3 The Properties of LSE

Without proof, we list below some results associated with least squares estimation, which are commonly used in the book. These properties collectively emphasize the significance of the *Least Squares Estimation* method in linear regression, contributing to its widespread application in financial statistical analysis and modeling.

1. *Unbiasedness*: LSE provides unbiased estimators for the regression coefficients β. This means that on average, the estimated coefficients $\hat{\beta}$ converge to the true population parameters as more data is collected. Theoretically, the LSE $\hat{\beta}$ is unbiased since $E(\hat{\beta}) = \beta$ with variance $var(\hat{\beta}) = (X'X)^{-1}\sigma^2$ if $var(\epsilon) = \sigma^2 I$. Furthermore, if the error term ϵ in equation (2.1) is normally distributed, such as $\epsilon \sim N(0, \sigma^2)$, it can be easily seen that $\hat{\beta} \sim N\left(\beta, (X'X)^{-1}\sigma^2\right)$.

This result supports the validity of the statistical inference with confidence intervals and hypothesis tests of *LSE* for the regression parameters using the t-test when the estimated $\hat{\sigma}^2$ is used in the statistical inference.

2. *Efficiency*: Among the linear unbiased estimators, *LSE* has the lowest variance. This property makes it an efficient estimator, producing reliable and stable coefficient estimates. This is also the property of *Best Linear Unbiased Estimators (BLUE)*, which signifies that, among all possible linear unbiased estimators, *LSE* offers the smallest variance, making it an optimal choice.

3. *Consistency*: As the sample size grows, *LSE* estimates converge to the true population parameters. This property of *LSE* is known as consistency.

4. *Interpretability*: *LSE* coefficients have straightforward interpretations. They represent the change in the response variable for a one-unit change in the predictor variable, keeping other predictors constant.

5. *The predicted values*: The predicted values are calculated as $\hat{y} = X\hat{\beta} = X(X'X)^{-1}X'y = Hy$, where $H = X(X'X)^{-1}X'$ is called the *hat-matrix* to turn observed y into *hat* $\hat{y}$.

6. *The residuals for diagnostics*: The regression residuals are defined as $\hat{\epsilon} = y - \hat{y} = y - X\hat{\beta} = (I - H)y$, which are the key components for model diagnostics.

7. *Residual sum of squares (RSS)*: The RSS is defined as $RSS = \hat{\epsilon}'\hat{\epsilon} = y'(I - H)'(I - H)y = y'(I - H)y$, which is used to estimate the residual variance and to evaluate the goodness of model fitting.

8. *Variance estimate*: It can be shown that $E(\hat{\epsilon}'\hat{\epsilon}) = \sigma^2(n - K - 1)$ (where $K + 1$ is the number of columns of the design matrix X, i.e., the number of parameters in the linear model). We then *estimate* σ^2 using $\hat{\sigma}^2 = \frac{\hat{\epsilon}'\hat{\epsilon}}{n-K-1} = \frac{RSS}{n-K-1}$ where $n - K - 1$ is the *degrees of freedom* of the model.

9. R^2: R^2 is called the *coefficient of determination* or percentage of response variation explained by the systematic component (i.e.,X, which is usually used as a goodness-of-fit measure): $R^2 = \frac{\sum(\hat{y}_i - \bar{y})^2}{\sum(y_i - \bar{y})^2} = 1 - \frac{\sum(y_i - \hat{y}_i)^2}{\sum(y_i - \bar{y})^2}$. R^2 ranges from 0 to 1, with values closer to 1 indicating better a better fit of the model.

2.2.4 The Fundamental Assumptions

Linear regression relies on several assumptions to ensure the validity and reliability of its results. These assumptions provide a framework for interpreting regression coefficients, making predictions, and conducting hypothesis tests in financial data analysis. Here are the key assumptions of linear regression:

1. *Linearity*: The relationship between the independent variable(s) and the dependent variable should be linear. This means that changes in the independent variable(s) should result in proportional changes in the dependent variable. This assumption can be validated by plotting the observed data to see whether there exists a linear relationship between the dependent variable and all other independent variables as illustrated in Figure 2.1.
2. *Independence of Errors*: The errors (the differences between observed and predicted values) should be independent of each other. *Autocorrelation* or *serial correlation* in residuals can lead to biased coefficient estimates and invalid hypothesis tests and we will discuss how to address this autocorrelation in Chapter 7 and Chapter 8.
3. *Homoscedasticity*: The variance σ^2 of the residuals should remain constant across all levels of the independent variable(s). *Heteroscedasticity* (unequal variance) can affect the efficiency of coefficient estimates and the validity of hypothesis tests.
4. *Normality of Residuals*: The residuals should follow a normal distribution. Departure from normality can impact the reliability of confidence intervals and hypothesis tests, particularly for smaller sample sizes.
5. *No Multicollinearity*: In *multiple linear regression*, the independent variables should not be highly correlated with each other. *Multicollinearity* can make it difficult to isolate the individual effects of predictors on the dependent variable.
6. *No Perfect Multicollinearity*: In multiple linear regression, there should be no perfect linear relationship between the independent variables. Perfect *multicollinearity* makes it impossible to determine the unique effect of each predictor.
7. *No Endogeneity*: The independent variables should not be correlated with the error term. *Endogeneity* can lead to biased and inconsistent coefficient estimates.
8. *No Outliers*: Outliers can disproportionately influence the regression model, affecting the coefficients and overall model fit. They should be identified and addressed appropriately.

9. *Independence of Observations*: The observations should be independent of each other. This assumption is particularly important in time series data, where *autocorrelation* may exist.

It's essential to assess these assumptions before interpreting the results of a linear regression for financial data analysis. Violations of these assumptions can lead to inaccurate and unreliable conclusions in financial recommendation. Various diagnostic tools and techniques are available to check these assumptions and take appropriate corrective measures if needed. We will examine these assumptions in the following data analyses.

2.3 Simple Linear Regression: One Independent Variable

As seen in Figure 2.1, there is a distinctive increasing trend between the *wage* and the *consumer price index(CPI)*. To examine this trend numerically, we can make use of a *simple linear regression*.

2.3.1 Model Fitting

The implementation of linear regression in *R* is very straightforward. The *R* function **lm** (i.e., linear model) will allow you to perform this linear regression.

For example, the simple linear regression (*SLM*) model on w over *CPI* can be done as follows:

```
# Simple linear regression
slm1 = lm(w ~ CPI,nwages)
```

We can print the model summary as follows. The model summary provides information on model fit (overall F-test for significance, R^2, and standard error of the estimate), parameter significance tests, and a summary of residual statistics.

```
# Print the model fitting
summary(slm1)
```

```
##
##
## Call:
## lm(formula = w ~ CPI, data = nwages)
##
## Coefficients:
```

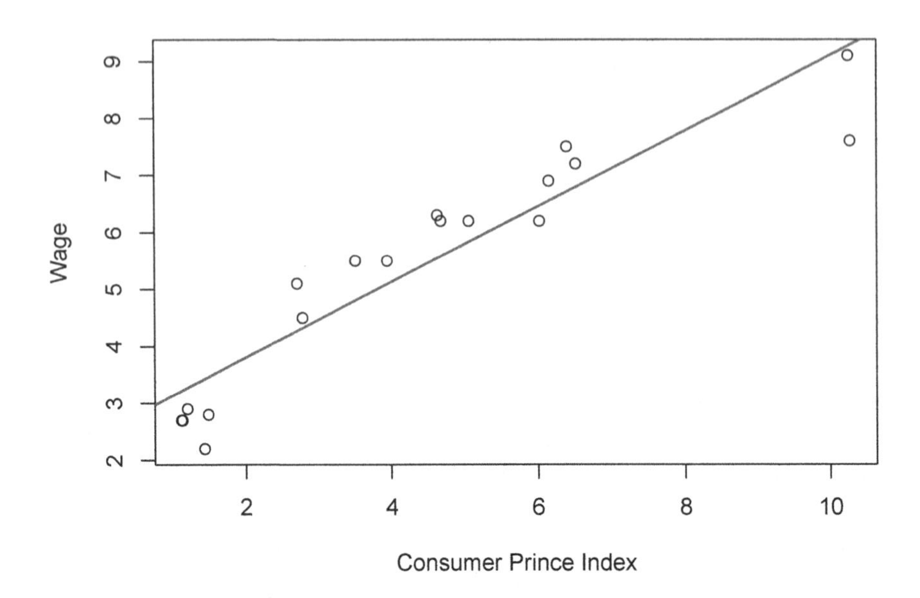

FIGURE 2.2
Illustration of Simple Linear Regression Fitting

```
##              Estimate Std. Error t value Pr(>|t|)
## (Intercept)    2.4821     0.3293    7.54  1.2e-06 ***
## CPI            0.6620     0.0633   10.46  1.5e-08 ***
##
## Residual standard error: 0.747 on 16 degrees of freedom
## Multiple R-squared:  0.872,  Adjusted R-squared:  0.864
## F-statistic:  109 on 1 and 16 DF,  p-value: 1.46e-08
```

As seen from this model summary, we obtained a significant simple linear regression model as seen from significant *F-statistic: 109.4 on 1 and 16 DF* which produced a *p-value: 1.461e-08*. From this simple linear regression model, the estimated slope parameter $\hat{\beta}_1 = 0.66199$ with standard error of 0.06328 gave *p-value = 1.46e-08*. This indicated that the *wage* is statistically significantly related with *consumer price index CPI* with one-unit change in *CPI* resulted in 0.66199 united of *wage* increase.

The estimated residual standard error $\hat{\sigma} = 0.7465$ and the estimated R^2 from this simple linear regression is 0.8724, which is quite satisfactory. This means that this significant regression model explains about 87.24% of the variation in the data. This can be seen in Figure 2.2, where we add the model fitting to the original data.

The programming language *R* operates in an *object-oriented* manner. Consequently, the results of a model fit are stored as an *R* object. This enables us to retrieve a wealth of supplementary information generated by the model, including predicted values and residuals. By utilizing the *attributes()* function within *R*, we can acquire a comprehensive list of the supplementary information accessible through the *lm* function as shown below:

```
# Print the contents of the "slm1" R object
attributes(slm1)
```

```
## $names
##  [1] "coefficients"  "residuals"     "effects"
##  [4] "rank"          "fitted.values" "assign"
##  [7] "qr"            "df.residual"   "xlevels"
## [10] "call"          "terms"         "model"
```

For example, to access the LSE parameter estimates, we can call the *coefficients* as follows:

```
# Show the parameter estimates
slm1$coefficients
```

```
## (Intercept)         CPI
##       2.482       0.662
```

which would give you the estimated parameters. Interested readers should get familiar with this trick since we will access *R* objects in this book all the time for data analysis and plotting.

2.3.2 Model Diagnostics

As outlined in Section 2.2.4, model assumption should be validated. For this purpose, we do model diagnostics on residuals and we can make use of *R* package *car* (i.e., *C*ompanion to *A*pplied *R*egression). Let's load this package first:

```
# Load the "car" package
library(car)
```

The function *residualPlots* in this package can be called to produce a *Tukey* test on residual homogeneity along with the residual plots as shown in Figure 2.3 with following *R* code chunk:

```
# Residual homogeneity plot and test
residualPlots(slm1)
```

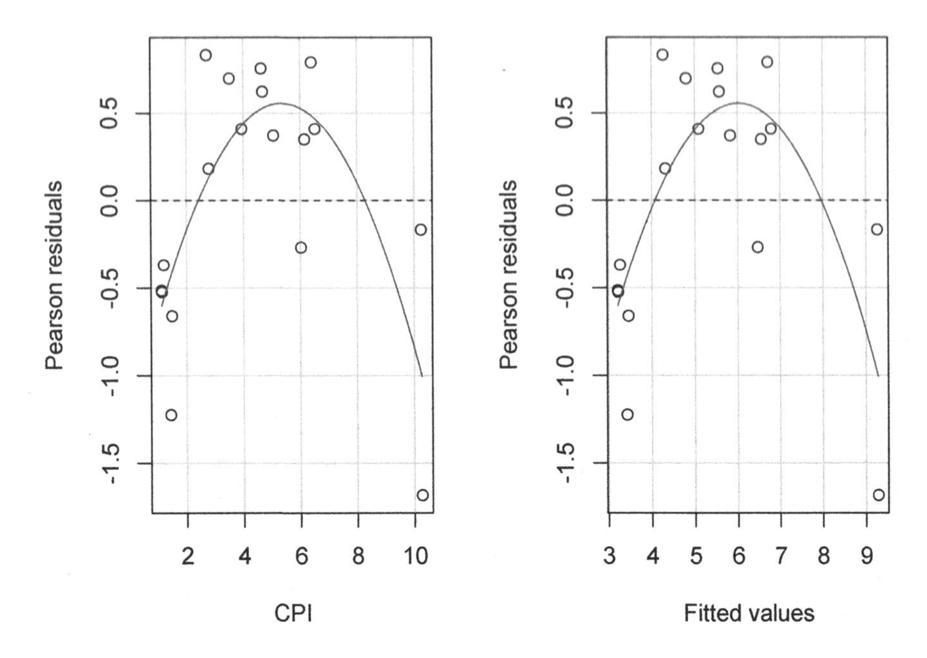

FIGURE 2.3
Illustration of Residual Plots

```
##             Test stat Pr(>|Test stat|)
## CPI             -4.96          0.00017 ***
## Tukey test      -4.96          7.1e-07 ***
## ---
```

The *p-values* from the *Tukey* test are both highly significant, thus indicating that there is a deviation from residual *homogeneity* and there is a need to include more variables into this *slm1* to improve the model fitting. However, the *QQ-plot* in Figure 2.4 indicated no obvious pattern of deviation from the *residual normality.*

```
# Residual qqnorm plot
qqPlot(slm1)
```

2.4 Multiple Linear Regression

Due to the deviation of the residuals from the simple linear regression above, let us add more variables for a *multiple linear regression* following the regression

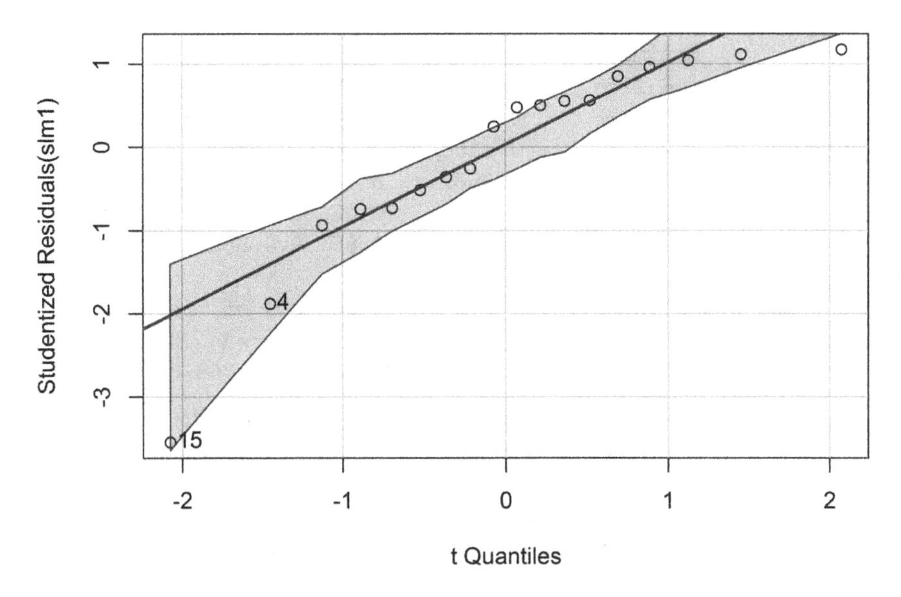

FIGURE 2.4
Illustration of Residual QQ-Plot

equation (6.10) at Nicols (1983) to examine the *wage* with other variables of *CPI1, CPI2, u, mw*. The *simple linear regression* now becomes *multiple linear regression* with these added variables.

2.4.1 Model Fitting

To implement *multiple linear regression* in *R*, we can simply add more variables into the *lm* function as follows:

```
# Nicols' multiple linear regression model
modelNichols <- w ~ CPI + CPI1 + CPI2 + u + mw
# Call lm to fit Nicols model
MLMNicols <- lm(modelNichols, data = nwages)
# Print the model summary
summary(MLMNicols)
```

```
## Call: lm(formula = modelNichols, data = nwages)
##
## Residuals:
##      Min       1Q  Median       3Q      Max
## -0.7401 -0.2103 -0.0224  0.2040  0.8699
##
```

```
## Coefficients:
##             Estimate Std. Error t value Pr(>|t|)
## (Intercept)   4.2754     0.4872    8.78  1.4e-06 ***
## CPI           0.5188     0.0860    6.04  5.9e-05 ***
## CPI1          0.1213     0.1166    1.04  0.31867
## CPI2          0.2140     0.0589    3.64  0.00341 **
## u            -0.4879     0.1103   -4.42  0.00083 ***
## mw            0.0316     0.0171    1.85  0.08883 .
## ---
## Residual standard error: 0.419 on 12 degrees of freedom
## Multiple R-squared:  0.97,	Adjusted R-squared:  0.957
## F-statistic: 77.2 on 5 and 12 DF,  p-value: 1.06e-08
```

This *multiple linear regression* is significant and the associated R^2 increased to 0.9699 from 0.8724 in the previous *simple linear regression*. With this *multiple linear regression* model, the *CPI* is still statistically significant along with the lagged-2 *CPI2*. The unemployment u is also statistically significant with the estimated slope parameter $\hat{\beta}_4$ = -0.487, indicating that the higher the unemployment, the lower the wage. The lagged-1 *CPI1* and the minimal wage *mw* are not statistically significantly associated with the wage w.

We can also use the *F-test* for the overall model from the entire ANOVA result including sums of squares and mean squares by using the *anova* function as follows:

```
# Print the anova table for F-test
anova(MLMNicols)
```

```
##
## Analysis of Variance Table
##
## Response: w
##           Df Sum Sq Mean Sq F value  Pr(>F)
## CPI        1   61.0    61.0  347.33 3.2e-10 ***
## CPI1       1    0.1     0.1    0.67  0.4306
## CPI2       1    3.3     3.3   18.51  0.0010 **
## u          1    2.8     2.8   16.17  0.0017 **
## mw         1    0.6     0.6    3.43  0.0888 .
## Residuals 12    2.1     0.2
##
```

A better presentation of the parameter estimation can be produced by using *R* package *xtable* as follows:

TABLE 2.1
Summary of the Parameter Estimation

	Estimate	Std. Error	t value	Pr(>|t|)
(Intercept)	4.2754	0.4872	8.776	0.0000
CPI	0.5188	0.0860	6.036	0.0001
CPI1	0.1213	0.1166	1.040	0.3187
CPI2	0.2140	0.0589	3.637	0.0034
u	-0.4879	0.1103	-4.424	0.0008
mw	0.0316	0.0171	1.852	0.0888

```
# Load the library
library(xtable)
# Make the table
knitr::kable(round(xtable(MLMNicols),4),
      caption = 'Summary of the Parameter Estimation',
      booktabs = TRUE )
```

As seen in Table 2.1, the variables of *CPI*, *CPI3*, and *u* are statistically significantly related to *wage*.

2.4.2 Regression Model Diagnostics: Checking Assumptions

The basic assumptions for *linear regression* are *normality*, *homogeneity*, and *independence*. Assumptions should be checked and validated for each data analysis. We illustrate the model diagnostics using the *multiple linear regression* model *MLMNicols*.

2.4.2.1 Normality

For residual *normality*, we would extract the regression residuals and then examine the distribution of the residuals using *hist* as follows:

```
# Histogram plot to show normality
hist(MLMNicols$residuals, main="Residual Distribution",
     xlab="Residuals from MLMNicols")
```

Figure 2.5 shows an approximately well-behaved one-mode distribution with an approximate *bell* shape due to the small number of observations.

We can also use the *QQ-plot* to plot the quantiles of the residuals to the theoretical normal quantiles to see whether there is a one-to-one relationship. In QQ-plot, the y-axis is for the quantiles from the observed residuals and the

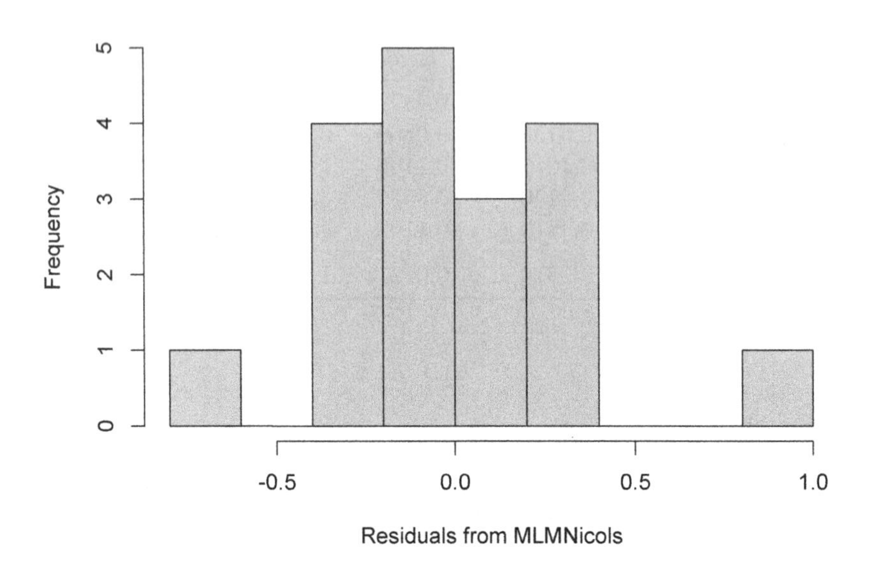

FIGURE 2.5
Residual Histogram Plot

x-axis is for the corresponding points of a normal distribution. The observed data (i.e. residuals) points in quantiles are plotted to the theoretical normal quantiles by black circles. The solid line represents the data conforming perfectly to the normal distribution. Therefore, the closer the observed data in black circles are to the solid line, the more closely the data conforms to the normal distribution.

There are two ways to show the QQ-plot. The first way is to use the default *stat* package and extract the residuals from the *lm* object. Then we can use *R* functions of *qqnorm* and *qqline* to make this plot as shown in Figure 2.6 with the following *R* code chunk:

```
# QQ-plot with residuals extracted from *MLMNicols*
qqnorm(MLMNicols$resid)
qqline(MLMNicols$resid)
```

The second way is to use *car* package which will show the associated 95% confidence band for the line. When data points fall within the 95% confidence band, they are considered to conform to the normal distribution. We call the *qqPlot* to make this QQ-plot as follows:

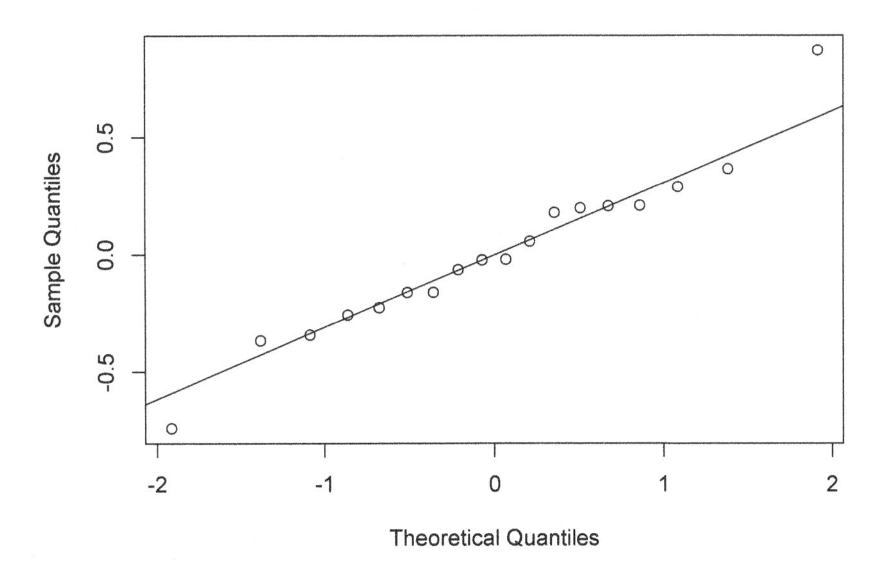

FIGURE 2.6
Residual QQ-Plot with Default *stat* Package

```
# QQ-plot with car  package
qqPlot(MLMNicols, distribution="norm")
```

From the *histogram* in Figure 2.5, the distribution of the residuals are symmetric and seems normally distributed. From the QQ-plot in Figure 2.7, we can see that most of the quantile points are close to the one-to-one line and within the 95% confidence bands except a few points at both ends.

If a statistical test is needed, we can call the *shapiro.test* to perform the *Shapiro-Wilk test of normality* as follows:

```
## Shapiro test for normality
shapiro.test(MLMNicols$residuals)
```

```
##
##  Shapiro-Wilk normality test
##
## data:  MLMNicols$residuals
## W = 0.97, p-value = 0.7
```

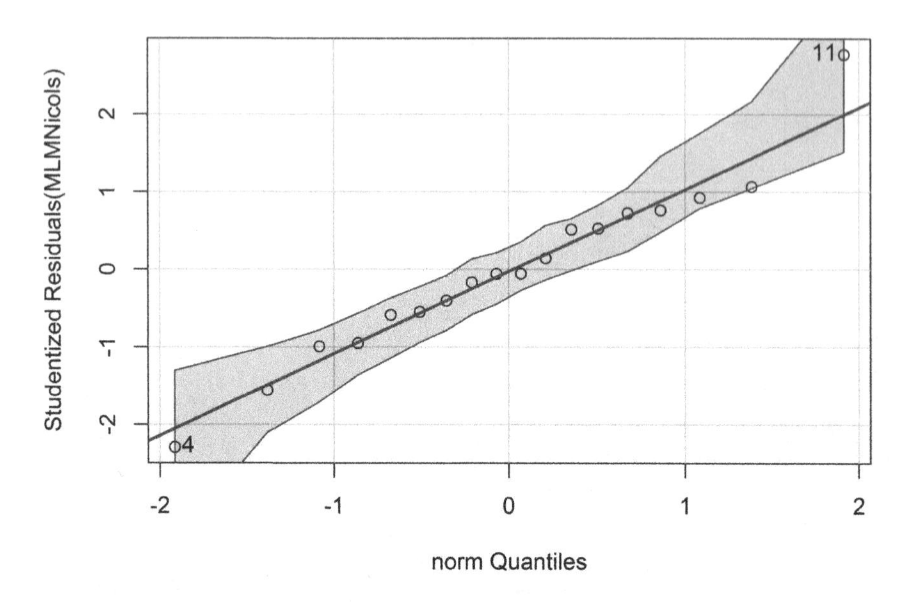

FIGURE 2.7
Residual QQ-Plot from *car* Package

The *p-value* associated with the Shapiro-Wilk test is 0.702, indicating that we do not reject the null hypothesis that the residuals from model *MLMNicol* are normally distributed.

2.4.2.2 Homogeneity

The residual plot is typically used for *homogeneity test* (i.e. constant variance). In *car* package, this is implemented in the function *residualPlots.* This function plots the residuals versus each variable and versus fitted values in the fitted model *MLMNicols.* In addition, this function gives a curvature test for each of the plots by adding a quadratic term and testing the quadratic term to be zero. For a *linear regression* model, this is *Tukey's test for nonadditivity* when plotting against fitted values.

Let's call this function for the model *MLMNicols*:

```
# Residual plot
residualPlots(MLMNicols,layout=c(2,3))
```

```
##            Test stat Pr(>|Test stat|)
## CPI            -1.89            0.085 .
## CPI1           -3.07            0.011 *
## CPI2           -2.31            0.041 *
```

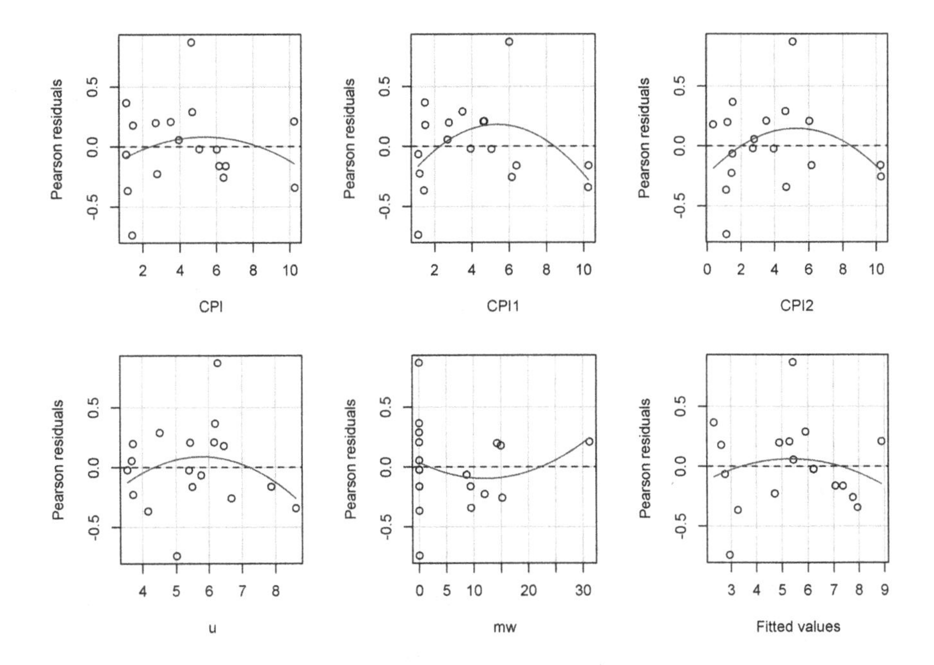

FIGURE 2.8
Residual Plot to Test Homogeneity

```
## u              -1.53             0.153
## mw              1.07             0.307
## Tukey test     -1.38             0.168
```

As seen from the output, no quadratic terms are needed for *CPI, u, mw,* but needed for *CPI1, CPI2* as seen in Figure 2.8. However, the overall *Tukey test* gave a satisfactory p-value of 0.16842.

2.4.2.3 Independence

Another assumption in regression is that the residuals/data should be independent. For time series data, there are usually temporally correlated with time series *autocorrelation.* We can check the residual time series autocorrelation with the *Durbin-Watson (DW)* test for autocorrelation as follows.

```
# Perform DW test
dwtest(MLMNicols)
```

```
##
##  Durbin-Watson test
##
```

```
## data:  MLMNicols
## DW = 1.3, p-value = 0.009
## alternative hypothesis:true autocorrelation is greater than 0
```

The *Durbin-Watson (DW)* test has the null hypothesis that the autocorrelation of the residuals is 0. Under the assumption of normally distributed residuals, the null distribution of the *Durbin-Watson* statistic is the distribution of a linear combination of chi-squared variables. The *p-value* is computed using a normal approximation with mean and variance of the *Durbin-Watson* test statistic. As seen in this data, the *DW = 1.2561, p-value = 0.009077*, indicating there is statistically significant time series autocorrelation in the residuals of model fitting *MLMNocols*. This is not surprising for time series data in financial data analysis. We will discuss this in Chapter 7.

2.5 Model Diagnostics with Monte-Carlo Simulations

As briefly discussed in Section 2.2.4, the validity of the linear regression relies on several assumptions to ensure the correct parameter estimation as well as the associated statistical inference and predictions in financial data analysis. Nine key assumptions were listed and we will design some simulation studies to demonstrate the consequences of violations of some of these assumptions focusing on the assumptions of *Normality of Residuals*, *Homoscedasticity*, and *Independence of Observations* in this chapter and we will leave the rest model diagnostics to other chapters.

This simulation exercise not only provides valuable insights and knowledge about linear regression but also serves as a pivotal starting point for interested readers to delve deeper into model diagnostics using Monte-Carlo simulations.

2.5.1 Violation of *Normality of Residuals*

As written in equation (2.2), the residual term (or error term) ϵ should be normally distributed with standard deviation of σ for correct statistical inference including the F-test for the model fit and the t-test for the parameter significance. Normal distribution is a symmetric and *bell*-shaped distribution. We will investigate the consequences if the ϵ is t-distributed with different degrees of freedom to model the heavy-tail influence of parameter estimation.

2.5.1.1 The t-Distrution v.s. The Normal Distribution

The t-distribution is a very population distribution. Like the normal distribution, it is bell-shaped and symmetric. However, unlike normal distributions, it has heavier tails, which can be used to model financial data with a greater chance for extreme values characterized by the associated degrees of freedom.

If ϵ is t-distributed with ν degrees of freedom, the probability density function (PDF) is as follows:

$$f(\epsilon) = \frac{\Gamma\left(\frac{\nu+1}{2}\right)}{\sqrt{\nu\pi}\Gamma\left(\frac{\nu}{2}\right)}\left(1+\frac{\epsilon^2}{\nu}\right)^{-\frac{\nu+1}{2}}. \tag{2.8}$$

With some mathematical manipulations, it is easy to show that the associated mean and variance of t-distribution is 0 and $\frac{\nu}{\nu-2}$, respectively. If ν is very small, the t-distribution will behave very heavy-tailed with special case of $\nu = 1$ to be the so-called *Cauchy*-distribution which is a symmetric distribution with undefined both mean and variance and PDF of $f(\epsilon) = \frac{1}{\pi(1+\epsilon^2)}$. As ν increases and approaches to infinite, the t-distribution becomes the standard normal distribution with PDF as follows:

$$f(\epsilon) = \frac{1}{\sqrt{2\pi}}e^{-\frac{\epsilon^2}{2}}. \tag{2.9}$$

Therefore, the degrees of freedom parameter ν characterizes any data from heavy-tailed (small ν) to normally distributed (large ν) in financial data analysis.

Let's simulate data from t-distribution and graphically display their distributions to see the property of degrees of freedom. *R* is built for data simulation. For t-distribution, we can make use *rt(n, df, ncp)* function to simulate t-distributed data. In this function, the n is the number of data points to be simulated, *df* is the degrees of freedom ν, and *ncp* is the non-centrality parameter which is zero to represent the central t-distribution in equation (2.9). The following *R* code chunk is to simulate two datasets with sample size $n = 500$ for the error term ϵ, where *Case 1* is for the t-distribution at 3 degrees of freedom to mimic the heavy-tailed t-distribution, and *Case 2* for the t-distribution at 100 degrees of freedom to mimic the standard normal distribution.

```
# Set the simulation seed for reproducibility
set.seed(3388)
# Let's simulate n = 500 observations
n=500
# Case 1: nu=3
nu = 3
ept1 = rt(n, df = nu)
# print the first 8 data points
ept1[1:8]
```

```
## [1]  1.9554  1.3114  0.1153  1.5265  0.2254 -0.7476
## [7] -0.1168  0.6235
```

```
# Case 2: nu=100
nu =100
ept2 = rt(n, df = nu)
# print the first 8 data points
ept2[1:8]
```

```
## [1] -0.6995 -1.0283 -0.5571 -0.9384 -1.3475  0.2948
## [7] -0.4341 -0.7740
```

With the generated data *ept1* and *ept2*, we can examine their distributions by plotting these distributions with an overlay of the standard normal density curve to see how different they are to the standard normal distribution. For this purpose, we can make use of the *R* package *ggplot2*, which is a wonderful *R* package for professional graphics. The steps are as follows:

1. *Step 1 to Create Histogram or Density Plot*: We can use either a *histogram* or a *density* plot to visualize the data distribution. With *ggplot2*, we use a density plot with *geom_density()* function.
2. *Step 2 to Overlay Normal Density*: We then overlay the standard normal density curve on the plot using the *stat_function()* function from *ggplot2*. This function allows us to add a statistical function to the plot, in this case, the *standard normal density function.*
3. *Step 3 to Put the Two Plots Side-by-Side*: There are two plots created for *ept1* and *ept2* by *Step 1* and *Step 2*. The third step is to put them side-by-side for visual display and comparison. This can be done using the *grid.arrange()* function from the *gridExtra* package.

The exact *R* coding for these 3 steps is as follows:

```
# Steps 1 and 2 to create plot: p1
#   with normal density curve overlaid
# Call library: ggplot2
library(ggplot2)
# Call ggplot to make plot p1
p1 = ggplot(data.frame(x=ept1), aes(x = ept1)) +
# Generate density plot for ept1
  geom_density(fill="lightblue",color = "black") +
# Add x-axis limit from -4 to 4
  scale_x_continuous(limits = c(-4,4)) +
# Overlay the standard normal density
```

```
  stat_function(fun=dnorm,args=list(mean=0,sd=1),
                    lwd=2,color="red") +
# Add a labs and rename the x-axis label
  labs(title = "Data Distribution with Normal Density Overlay",
       y="", x = "t-Distribution with Degrees of Freedom of 3")
# Steps 1 and 2 to create plot: p2
#   with normal density curve overlaid
p2 = ggplot(data.frame(x=ept2), aes(x = ept2)) +
  geom_density(fill = "lightblue", color = "black") +
  scale_x_continuous(limits = c(-4,4)) +
  stat_function(fun = dnorm, args = list(mean = 0, sd = 1),
                lwd=2, color = "red") +
        labs(title = "", y="",
       x = "t-Distribution with Degrees of Freedom of 100")
# Step 3 to arrange the plots side by side
# Call library gridExtra
library(gridExtra)
# Call grid.arrange function to put p1 and p2 together
grid.arrange(p1, p2, ncol = 2)
```

Not surprised as seen from Figure 2.9, the right-side plot *p2* is very close to the standard normal distribution due to the large degrees of freedom whereas the left-side plot *p1* is heavy-tail due to the small degrees of freedom of 3.

2.5.1.2 Example of Impact with One Simulation

To investigate the impact of *nonnormal residual* distribution, Let's simulate data from a regression which we know the truth. We make use of a *simple linear regression* with only one independent variable x generated from the standard normal distribution. The true parameters are $\beta_0 = 1$ and $\beta_1 = 1$ with the error variance $\sigma = 1$, such as:

$$y = \beta_0 + \beta_1 \times x + \sigma \times \epsilon \tag{2.10}$$

where the error term ϵ follows the t-distribution with the degrees of freedom of 3 and 100.

With the above simple linear regression (*SLR*) model, we can get the parameter estimates $(\hat{\beta}_0, \hat{\beta}_1, \hat{\sigma}^2)$ using *R* function *lm* and the associated variances can be calculated using the formula in Assumption 1 from Section 2.2.4, which is $var(\hat{\beta}) = (X'X)^{-1}\hat{\sigma}^2$.

The data generation and the estimation can be coded and explained in the following *R* code chunk:

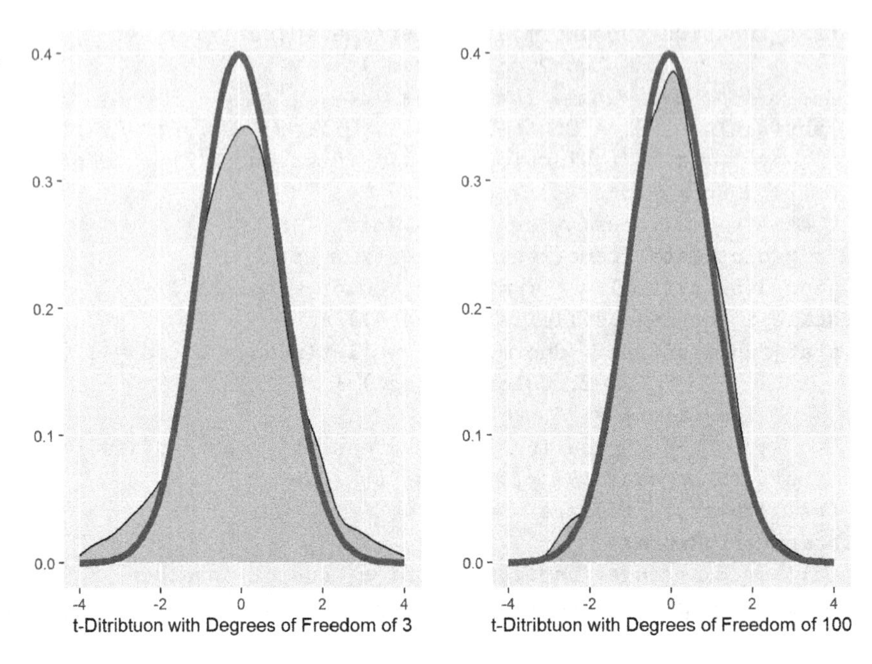

FIGURE 2.9
Illustration of t-Distribution Overlaid by Normal Distribution

```
# Set the random seed
set.seed(3388)
# True parameters
b0=1; b1=1; sigma=1
# Generate x from standard normal
x = rnorm(n)
# Calculate the ys from both cases
y1 = b0+b1*x+sigma*ept1;
y2 = b0+b1*x+sigma*ept2
# With the data generated, we can use fit linear models
modt1 = lm(y1~x); modt2 = lm(y2~x)
# Extract the estimated parameters from Case 1
est1 = coef(modt1);
# Print the estimates
cat("Estimated Parameters in Case 1:",
    as.numeric(est1),"\n")
```

```
## Estimated Parameters in Case 1: 1.028 1.129
```

```
estsig1 = summary(modt1)$sigma;
cat("Estimated Sigma in Case 1:",
    as.numeric(estsig1),"\n")

## Estimated Sigma in Case 1: 1.725

# Extract the estimated parameters from Case 2
est2 =coef(modt2);
cat("Estimated Parameters in Case 2:",
    as.numeric(est2),"\n")

## Estimated Parameters in Case 2: 1.036 0.9936

estsig2 = summary(modt2)$sigma;
cat("Estimated Sigma in Case 2:",
    as.numeric(estsig2),"\n")

## Estimated Sigma in Case 2: 1.022

# Calculate the estimated variance of \hat\beta
Xmat = model.matrix(modt1)
est1.var = diag(solve(t(Xmat)%*%Xmat)*estsig1^2);
cat("Estimated Variance for Parameters in Case 1:",
    as.numeric(est1.var),"\n")

## Estimated Variance for Parameters in Case 1: 0.005949 0.006079

est2.var = diag(solve(t(Xmat)%*%Xmat)*estsig2^2);
cat("Estimated Variance for Parameters in Case 2:",
    as.numeric(est2.var),"\n")

## Estimated Variance for Parameters in Case 2: 0.002088 0.002133

# Calculate the true variance
est.var = diag(solve(t(Xmat)%*%Xmat)*sigma^2);
cat("True Variance for Parameters:",
    as.numeric(est.var),"\n")

## True Variance for Parameters: 0.002 0.002044
```

As seen from this one simulation, the estimated parameters for *Case 1* are $(\hat{\beta}_0, \hat{\beta}_1, \hat{\sigma}^2) = (1.028346, 1.128795, 1.724637)$ and (1.035806, 0.9936139,

1.021629) for *Case 2*. In contrast with the truth of $(\beta_0, \beta_1, \sigma^2) = (1, 1, 1)$, the *Case 1* has the larger bias than those in *Case 2* for all the three parameters, especially the error variance σ^2.

Similarly, the estimated variances for the estimated regression parameters $(\hat{\beta}_0, \hat{\beta}_1) = (0.005949087, 0.006078828)$ for *Case 1* and (0.002087573, 0.0021331) for *Case 2* where the true variances for these parameters are (0.002000115, 0.002043735). This is evident that the estimated variances for *Case 1* are bigger than those in *Case 2* due to the heavy-tailed t-distribution in *Case 1* with 3 degrees of freedom.

2.5.1.3 Full-Scale Monte-Carlo Simulation Study

To fully investigate this bias in parameter estimation and the associated statistical inference, we need to carry out a full-scale simulation study with a large number of simulations (for example, 10,000 simulations) instead of one simulation. With this large number of simulations, we can then look at the sampling distributions of the estimated parameters to examine the biases and other statistical properties.

This can be implemented easily using *R* loops as follows:

```
# Set the number of simulations
nsimu=10000
# Create a temporary matrix to hold the simulated output \
estMat = matrix(0, ncol=10, nrow=nsimu)
colnames(estMat) = c("b0.1","b1.1","sig1.1",
                     "b0.2","b1.2","sig1.2",
                    "varb0.1","varb1.1",
                    "varb0.2","varb1.2")
# Now loop over to estimate the parameters
for(s in 1:nsimu){
# Get the x variable
x = rnorm(n)
# Get the error terms for both cases
ept1 = rt(n, df = 3); ept2 = rt(n, df = 100)
# Get the simulated y variable
y1   = b0+b1*x+ept1; y2 = b0+b1*x+ept2
# Fit the linear regression models
modt1 = lm(y1~x); modt2 =lm(y2~x)
# Extract the estimated parameters
est1 = coef(modt1); sig1 = summary(modt1)$sigma;
estMat[s, 1:3] = c(est1,sig1)
est2 =coef(modt2);sig2 = summary(modt2)$sigma
estMat[s,4:6] = c(est2, sig2)
# Calculate the estimated variance of \hat\beta
```

```
Xmat = model.matrix(modt1)
estMat[s,7:8] = diag(solve(t(Xmat)%*%Xmat)*estsig1^2);
estMat[s,9:10] = diag(solve(t(Xmat)%*%Xmat)*estsig2^2);
} #End of the looping
# Print the first 6 rows and round to 3-digit
round(head(estMat,6),4)
```

```
##          b0.1   b1.1 sig1.1   b0.2   b1.2 sig1.2 varb0.1
## [1,] 0.9149 0.9466  2.383 1.0550 0.9931 0.9799  0.0060
## [2,] 0.8232 0.9351  2.159 0.9426 0.9279 0.9980  0.0060
## [3,] 0.9590 1.0532  1.614 1.0206 0.9762 0.9984  0.0060
## [4,] 1.0094 1.0764  1.850 1.1123 1.0012 1.0352  0.0060
## [5,] 1.0492 0.9050  1.435 1.0231 1.0039 0.9977  0.0059
## [6,] 1.0062 0.8382  1.581 1.0116 0.9851 0.9586  0.0060
##      varb1.1 varb0.2 varb1.2
## [1,]  0.0055  0.0021  0.0019
## [2,]  0.0057  0.0021  0.0020
## [3,]  0.0056  0.0021  0.0020
## [4,]  0.0064  0.0021  0.0022
## [5,]  0.0062  0.0021  0.0022
## [6,]  0.0061  0.0021  0.0021
##
```

With these 10,000 simulations, we can calculate the sample means from these 10,000 samples using *R* function *apply* as follows:

```
apply(estMat,2,mean)
```

```
##     b0.1     b1.1   sig1.1     b0.2     b1.2   sig1.2
## 0.999330 1.000699 1.699460 1.000360 1.000162 1.009278
##  varb0.1  varb1.1  varb0.2  varb1.2
## 0.005961 0.005985 0.002092 0.002100
```

As seen from the above, the overall mean for the estimated $(\hat{\beta}_0, \hat{\beta}_1, \hat{\sigma}^2)$ = (0.997796285, 0.999049080, 1.703182006) for *Case 1* and (1.000045786, 1.001198568, 1.009684307) for *Case 2*, respectively, where the true parameters are $(\beta_0, \beta_1, \sigma^2) = (1,1,1)$. This shows that the estimated (β_0, β_1) are unbiased, but σ^2 is biased higher in *Case 1* due to the heavy-tailed t-distribution with DF = 3 where it is almost unbiased in *Case 2* since the large degrees of freedom (i.e., DF = 100). Furthermore, the estimated variances for $(\hat{\beta}_0, \hat{\beta}_1)$ are (0.005961049, 0.005996747) for *Case 1* and (0.002091771, 0.002104297) for *Case 2* in comparison to the true variances of (0.002000115, 0.002043735). It is evident again that the estimated variances in *Case 1* are larger than those in

Case 2 due to the heavy-tailed t-distribution. This evidence can be graphically shown in the following figure with *R* code chunk below:

```
par(mfrow=c(2,3))
# Make the plots for Case 1
hist(estMat[,1], xlim=c(0.7,1.3),
     nclass=20, main="", xlab="Case 1: Beta0")
abline(v=b0, lwd=2, col="red")
hist(estMat[,2],xlim=c(0.7,1.3),
     nclass=20,  main="", xlab="Case 1: Beta1")
abline(v=b1, lwd=2, col="red")
hist(estMat[,3], xlim=c(0.7,2.5),
     nclass=30,  main="", xlab="Case 1: Sigma")
abline(v=sigma, lwd=2, col="red")

# Make the plots for Case 2
hist(estMat[,4], xlim=c(0.7,1.3),
     nclass=20, main="", xlab="Case 2: Beta0")
abline(v=b0, lwd=2, col="red")
hist(estMat[,5],xlim=c(0.7,1.3),
     nclass=20,  main="", xlab="Case 2: Beta1")
abline(v=b1, lwd=2, col="red")
hist(estMat[,6], xlim=c(0.7,2.5),
     nclass=30,  main="", xlab="Case 2: Sigma")
abline(v=sigma, lwd=2, col="red")
```

As seen from Figure 2.10, even though the estimations for (β_0, β_1) from these Monte-Carlo sampling distributions are unbiased, the estimation for σ is biased for *Case 1* as seen from the last plot in the first row.

In summary, the conclusions of this Monte-Carlo simulation study underscore a crucial observation: in situations where the underlying assumption of normally distributed residuals is compromised, the process of parameter estimation, particularly in relation to the error variance symbolized by σ, can incur bias. This bias, in turn, possesses the potential to instigate inaccuracies in the statistical inference, rendering the estimations of parameter variances unreliable.

In essence, this simulation study amplifies the significance of adhering to the normality assumption for residuals within linear regression models. By deviating from this fundamental assumption, not only does the validity of parameter estimation become compromised, but also the subsequent statistical inferences that stem from these estimations. Therefore, adherence to the pre-requisite assumptions, such as normality, is important for ensuring the validity

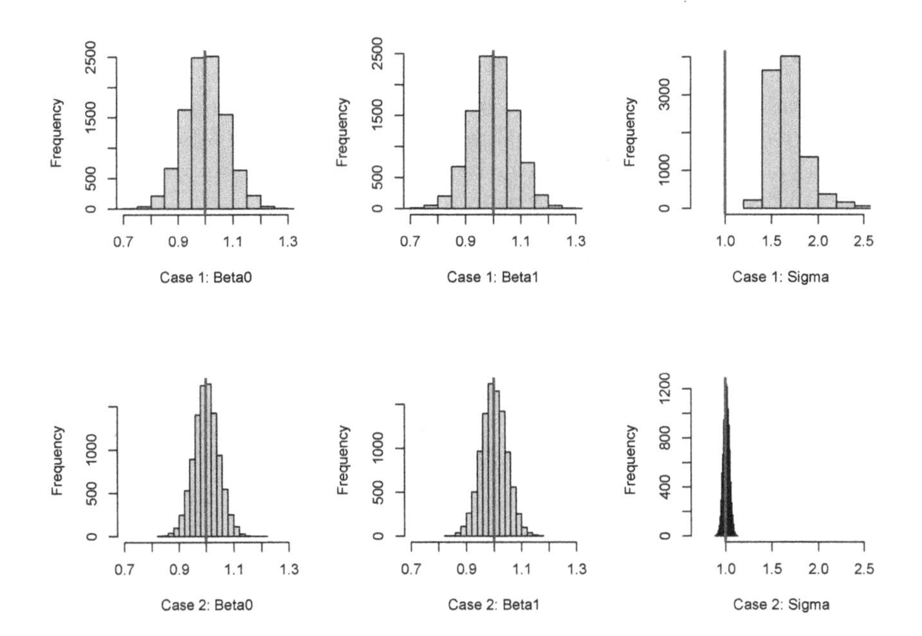

FIGURE 2.10
Monte-Carlo Distributions Overlaid with the True Parameters

and robustness of the analytical framework, thereby bolstering the credibility of the conclusions drawn from the financial data analysis.

2.5.2 Violation of *Homoscedasticity*

The violation of *homoscedasticity* within a statistical regression can have significant consequences, affecting the reliability and accuracy of the results in financial data analysis. *Homoscedasticity* refers to the assumption that the variance of the residuals (or errors) remains consistent across all levels of the predictor variables. When this assumption is violated, *heteroscedasticity* occurs. In the violation of *Homoscedasticity*, several negative consequences may arise in financial data analysis using regression modeling with the most common consequences as follows:

1. *Biased Coefficient Estimates*: In the presence of heteroscedasticity, the coefficient estimates in a linear regression model can become biased.

2. *Inaccurate Standard Errors*: The standard errors of the coefficient estimates are usually calculated under the assumption of homoscedasticity. In the presence of heteroscedasticity, these standard errors can

be inaccurate, leading to erroneous conclusions about the significance of predictors in the regression model.

3. *Invalid Hypothesis Tests*: Incorrect standard errors can lead to incorrect p-values in hypothesis tests. This can result in false positives (Type I errors) or false negatives (Type II errors), impacting the validity of statistical inferences.
4. *Misleading Confidence Intervals*: Confidence intervals provide a range within which the true parameter value is likely to lie. Heteroscedasticity can lead to wider or narrower confidence intervals that fail to accurately capture the parameter's true value.
5. *Inefficient Estimators*: Heteroscedasticity reduces the efficiency of coefficient estimates, making them less precise. Inefficient estimators lead to wider confidence intervals and less powerful hypothesis tests.
6. *Challenging Model Interpretation*: Interpretation of coefficients becomes more challenging in the presence of heteroscedasticity. It becomes difficult to discern the actual relationship between predictors and the response variable.
7. *Compromised Predictions*: When heteroscedasticity is present, the model's predictive accuracy may decrease. Predictions may have larger prediction intervals, making it harder to confidently predict outcomes.
8. *Limited Generalizability*: Models with heteroscedasticity may have limited generalizability to new or unseen data. The relationship learned from the current data might not hold when applied to different contexts.

In this section, we will investigate the *Misleading Confidence Intervals* that fail to accurately capture the parameter's true value in regression modeling. This will result *Inefficient Estimators* with incorrect confidence intervals and less powerful hypothesis tests with violation of *Homoscedasticity*. Interested readers can follow the logic and the *R* code developed in this section to investigate other scenarios.

2.5.2.1 From *Confidence Interval* to *Coverage Probability*

As a fundamental concept in statistics that helps quantify the uncertainty associated with point estimates, the *confidence interval (CI)* is a range of values within which the true regression parameters are estimated to lie with a specified level of confidence (typically 95% confidence interval). To evaluate whether this *CI* correctly covers the true parameters, *coverage probability (CP)*, often referred to as *confidence interval coverage probability*, is a concept in statistics that quantifies how often a calculated confidence interval captures the true population parameter it is intended to estimate. It provides a measure of the reliability and accuracy of the confidence intervals.

When we calculate a confidence interval, we are estimating a range within which we believe the true regression parameter lies with a certain level of confidence (such as 95%). The *coverage probability* indicates the percentage of times that such intervals, if repeatedly calculated from different samples, would contain the true parameter.

For example, if you calculate a 95% confidence interval for a regression parameter (say β_1), the coverage probability represents the proportion of times, in the long run, that the calculated interval would capture the true β_1. More formally, let's say we have a confidence interval with a stated confidence level of 95%. The *coverage probability* is the proportion of intervals calculated from different samples that would contain the true regression parameter. If the *coverage probability* is close to the stated confidence level (e.g., around 95% in this case), it indicates that the confidence intervals are reliable and accurately reflect the uncertainty associated with the parameter estimation.

Coverage probability is a critical aspect of assessing the quality of confidence intervals and the validity of statistical inference. If the *coverage probability* is consistently lower or higher than the stated confidence level, it suggests that the intervals are either too narrow (*undercoverage*) or too wide (*overcoverage*), and adjustments to the estimation procedure may be needed.

Therefore, to investigate the *Misleading Confidence Intervals* and *Inefficient Estimators* in the violation of *Homoscedasticity* when regression modeling is performed, we can examine the *coverage probability* from the resulted regression results to measure how well confidence intervals capture the true regression parameter values.

2.5.2.2 Example of Impact with One Simulation

To make it clear, we assume $\sigma = |x|$ to make sure σ is positive as shown in Figure 2.11. This would lead $\sigma^2 = x^2$ where σ^2 is not a constant anymore, but in relationship with the independent variable x where the *Homoscedasticity* is obviously violated. Be reminded that the $\sigma^2 = x^2$ is used to constrain that the σ^2 should be positive.

We still assume the true regression parameters (β_0, β_1)=(1,1) and investigate the 95% CI associated and the 95% coverage probability with these two regression parameters.

We will still utilize the simple *linear regression* model using the following *R* code chunk with detailed explanations:

```
# Set the random seed for reproducibility
set.seed(3388)
```

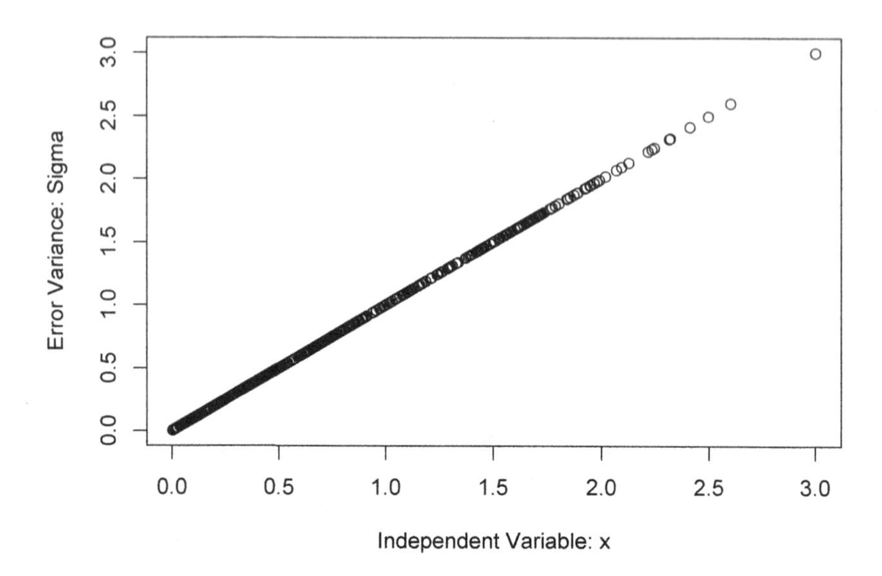

FIGURE 2.11
Heteroscedasticity: Relationship between the Sigma and x

```
# True parameters
b0=1; b1=1; true.par = c(b0,b1)
# Generate x from standard normal
x = rnorm(n)
# Generate the error term with sigma = abs(x)
sigma = abs(x); error = rnorm(n, mean=0, sd=sigma)
# Generate the y
y = b0+b1*x+error;
# With the data generated, we can fit linear models
modHomo = lm(y~x);
# Extract the estimated parameters
est.par = coef(modHomo);
# Extract the estimated sigma
est.sig = summary(modt2)$sigma
# Print the estimates
cat("Estimated Parameters:",
    c(as.numeric(est.par), est.sig),"\n")
```

```
## Estimated Parameters: 1.026 0.8483 0.992
```

```
# Calculate the estimated variance of \hat\beta
Xmat = model.matrix(modHomo)
estVar.par = diag(solve(t(Xmat)%*%Xmat)*est.sig^2)
cat("Estimated variance of parameters",
    estVar.par,"\n")

## Estimated variance of parameters 0.001968 0.002011

# Construct 95% CI: lower bound
CI.low = est.par-1.96*estVar.par;CI.low

## (Intercept)           x
##      1.0217      0.8444

# Construct 95% CI: upper bound
CI.up = est.par+1.96*estVar.par;CI.up

## (Intercept)           x
##      1.0295      0.8523

coverage = (true.par > CI.low) & (true.par < CI.up)
cat("Whether 95% CI covers the true parameter:",
    coverage,"\n\n")

## Whether 95% CI covers the true parameter: FALSE FALSE

# Also, plot the relationship between sigma and x in Figure 2.11
plot(sigma~abs(x), xlab="Independent Variable: x",
     ylab="Error Variance: Sigma")
```

As seen from this simulation, the estimated regression parameters are $\left(\hat{\beta}_0, \hat{\beta}_1\right)$ = (1.025601, 0.8483183) with estimated error standard deviation $\hat{\sigma} = 0.998482$. Therefore, the estimated variances $var(\hat{\beta}) = (X'X)^{-1}\hat{\sigma}^2 = (0.001994048, 0.002037535)$. With these values, we can construct 95% CI using formula $\hat{\beta} \pm \sqrt{var(\hat{\beta})}$ to get the lower bound as (1.0216925, 0.8443247) and upper bound as (1.0295092, 0.8523118) for $\beta = (\beta_0, \beta_1)$.

This is to say that the 95% CI for β_0 is (1.0216925, 1.0295092) and for β_1 is (0.8443247, 0.8523118). None of these two 95% CIs covered the true regression parameters $(\beta_0, \beta_1) = (1,1)$. Therefore, we have a case of *Misleading Confidence Intervals* in the violation of *Homoscedasticity* when regression modeling is performed where the resulted confidence intervals can not capture and cover the true regression parameter values.

2.5.2.3 Full-Scale Monte-Carlo Simulation Study

One case is not sufficient to demonstrate the *Misleading Confidence Intervals* and *Inefficient Estimators.* We can easily expand the one-case simulation to a full-scale Monte-Carlo simulation with a large number of simulations (for example, 10,000 simulations).

With this large number of simulations, we can then examine whether we can get correct 95% coverage probability for the regression CIs.

This can be implemented easily using *R* loops as follows:

```
# Set the number of simulations
set.seed(3388)
nsimu=10000
# Create a temporary matrix to hold the simulated output
covMatHomo = matrix(0, ncol=2, nrow=nsimu)
colnames(covMatHomo) = c("CP.b0","CP.b1")
# Now loop over to estimate the parameters
for(s in 1:nsimu){
    # Generate x from standard normal
    x = rnorm(n)
    # Generate the error term with sigma = abs(x)
    sigma = abs(x); error = rnorm(n, mean=0, sd=sigma)
    # Generate the y
    y = b0+b1*x+error;
    # With the data generated, we can use fit linear models
    modHomo = lm(y~x);
    # Extract the estimated parameters
    est.par = coef(modHomo);
    # Extract the estimated sigma
    est.sig = summary(modt2)$sigma
    # Calculate the estimated variance of \hat\beta
    Xmat = model.matrix(modHomo)
    estVar.par = diag(solve(t(Xmat)%*%Xmat)*est.sig^2)
    # Construct 95% CI: lower bound
    CI.low = est.par-1.96*estVar.par;CI.low
    # Construct 95% CI: upper bound
    CI.up = est.par+1.96*estVar.par;CI.up
    coverage = (true.par > CI.low) & (true.par < CI.up)
    covMatHomo[s, 1:2] = coverage;
} #End of the looping
```

```
# Print the first 10 rows
head(covMatHomo,10)
```

```
##       CP.b0 CP.b1
##  [1,]     0     0
##  [2,]     0     0
##  [3,]     1     0
##  [4,]     0     0
##  [5,]     0     0
##  [6,]     1     0
##  [7,]     0     0
##  [8,]     0     0
##  [9,]     0     0
## [10,]     0     0
```

As seen from this simulation, the *covMatHomo* is a matrix to hold whether the true parameter $\beta = (\beta_0, \beta_1)$ is covered by the 95% CIs with 1 denoting it is covered and 0 denoting that is not covered. We can then calculate the coverage probability by summing up how many of these 95% CIs covered the true regression parameters from these 10,000 samples using *R* function *apply* with the *sum* option as follows:

```
apply(covMatHomo,2,sum)
```

```
## CP.b0 CP.b1
##   672   389
```

As shown, we obtained only 672 and 389 of these 10,000 simulations which covered the true regression parameters with only 0.672% and 0.389% coverage probabilities for β_0 and β_1. These are far less than the required 95% which is the exact evidence of *Misleading Confidence Intervals* and *Inefficient Estimators.*

2.5.2.4 What to Do in Heteroscedasticity

When the assumption of homoscedasticity is violated in a linear regression analysis, there are several approaches we can take to address the issue and ensure the reliability of the regression analysis. Here are some tips we can recommend:

1. *Data Transformation*: Consider applying data transformations to stabilize the variance. Common transformations include the square-root, logarithm, or inverse-transformations. Experiment with different transformations to find one that improves the homoscedasticity of the residuals.

2. *Weighted Least Squares (WLS)*: In cases of heteroscedasticity, using *Weighted Least Squares (WLS)* can be beneficial. WLS assigns different weights to observations based on their variance, giving more weight to observations with lower variance. This approach helps account for the varying levels of variability in the data.

3. *Robust Regression*: Robust regression techniques, such as the Huber and Tukey estimators, are less sensitive to outliers and heteroscedasticity. These methods provide more robust coefficient estimates and standard errors.

4. *Segmentation*: If the *heteroscedasticity* is related to specific subsets of your data, consider segmenting the data and analyzing each subset separately. This can help address varying levels of variability in different segments.

5. *Include Additional Variables*: Including additional predictor variables that capture the sources of *heteroscedasticity* can sometimes help mitigate the issue. These variables can account for the systematic changes in variance across different levels of the predictors.

6. *Residual Analysis*: Perform a thorough analysis of the residuals to identify patterns that could indicate the source of *heteroscedasticity*. Plotting residuals against predictors or fitted values can provide insights into potential nonlinear relationships.

7. *Cluster-Robust Standard Errors*: If the *heteroscedasticity* is clustered by groups or clusters, you can use cluster-robust standard errors to adjust for the correlation within clusters while estimating standard errors.

8. *Nonparametric Methods*: Consider *nonparametric regression* methods, such as *kernel regression* or *local polynomial regression*, which can be more flexible in capturing the relationship between predictors and the response without assuming constant variance.

9. *Model Simplification*: Review your model and assess whether all included predictors are necessary. Simplifying the model by removing non-contributing predictors might help reduce *heteroscedasticity*.

10. *Sensitivity Analysis*: Perform sensitivity analyses by using different methods and assessing the robustness of your results across different approaches.

It's essential to choose an approach that aligns with the characteristics of your data and the objectives of your analysis. Therefore it is vital to understand your financial data with concrete financial theory in financial data analysis.

2.6 Discussions

In this chapter, we have done a thorough exploration of linear regression modeling. As a practical demonstration, we illustrated the regression techniques by employing a pertinent financial dataset known as *wages*. This dataset, initially introduced by Nicols (1983) and subsequently analyzed by Krämer and Sonnberger (1986), holds valuable insights into the dynamics of *wages* within the financial landscape. Notably, this dataset is readily accessible within the *R* library *lmtest*, which we conveniently employed to illustrate the various aspects of regression diagnostics.

To foster a clear understanding, we used a step-by-step approach utilizing the *R* programming language to analyze the *wages* dataset using both simple linear regression (*SLR*) and multiple linear regression (*MLR*) techniques. Our approach was designed to equip interested readers with a practical and in-depth comprehension of how these regression models can be employed to unravel relationships within financial data.

Careful attention should be paid to the validity of the assumptions underpinning linear regression modeling. These assumptions serve as the foundation upon which the model's accuracy and reliability lie. Conducting model diagnostics is a vital step in using linear regression to analyze financial data. This diagnostics process goes beyond mere assumption-checking; it involves critically evaluating the model's fit, residuals, influential points, multicollinearity, and other important model assumptions. We have demonstrated the consequences if these assumptions are violated using a Monte-Carlo simulation approach in this chapter. Interested readers can use the developed *R* code and associated ideas to validate the statistical analysis.

2.7 Exercises

1. Conduct Monte-Carlo simulation study for *Cauchy distribution* as mentioned in Section 2.5.1.1 since *Cauchy distribution* is a special t-distribution with the degree of freedom $\nu = 1$, but without finite mean and variance. This exercise is to see the impact of this distribution in *linear regression* even the *error distribution* has undefined mean and variance.

2. Conduct Monte-Carlo simulation study following the steps in Section 2.5.2 for variance function of $\sigma = \frac{1}{1+e^{-x}}$. This variance function is a function of the independent variable, but it is bounded.

3. This exercise is to investigate the effect of linear regression for correlated residuals, which violate the assumption of independence. For this purpose, use simple linear regression $y = \beta_0 + \beta_1 x + \sigma\epsilon$ with $(\beta_0, \beta_1, \sigma) = (1, 1, 1)$. ϵ is simulated from first-order autoregressive process with autocorrelations of 0.1 (small), 0.5(middle), 0.9 (high). Run these simulations for 10,000 times. Hint: ϵ can be simulated using *R* function *arima.sim(model = list(ar = rho), n = samplesize)* where *rho* is the autocorrelation coefficient which can be specified as 0.1, 0.5, or 0.9 in this exercise.

3

Transition from Linear to Nonlinear Regression

Not all relationships and models in finance are linear. The truth is that there are more nonlinear models than linear models that exist in real-life financial applications. In reality, the prevalence of nonlinear models substantially outweighs that of their linear counterparts within real-life financial applications. The complexities of financial systems, market behaviors, and economic phenomena often manifest in dynamic ways that a linear model can capture.

However, it has been noticed that many financial decisions can be better described using nonlinear models in practice. It is also known that financial managers and decision-makers are not very conversant with nonlinear models as compared to linear models, due to their simplicity. This chapter is then aimed to transform some of the nonlinear models commonly used in finance to their equivalent linear forms. The resulting linear models can then be modeled with the linear regression described in Chapter 2.

There are many nonlinear financial problems and it is impossible to include all of them into one chapter. In this chapter, we will select two of the most commonly-used nonlinear financial problems: the financial compounding model and the non-current asset depreciation model. These are typically exponential nonlinear models that can be transformed easily to corresponding linear models so that linear regression can be applied to estimate their parameters.

We will also take a further look at the pros and cons of linearization (i.e. transforming exponential nonlinear models into linear models), while focusing on the associated error distributions between them. To provide a clearer picture, we will conduct a Monte-Carlo simulation study, highlighting the need for careful consideration when adopting this practice. Through this exercise, we aim to provide a comprehensive understanding of the potential benefits and limitations associated with transforming these nonlinear models.

3.1 The Financial Compounding Model

We begin by introducing the fundamental concept of compounding, which stands as one of the cornerstones of financial modeling. At its core, compounding encompasses a powerful principle: the process of generating additional earnings or returns based on an initial investment. This is accomplished by reinvesting

DOI: 10.1201/9781003469704-3

previously earned or accrued gains back into the investment. In simpler terms, compounding allows you to not only earn returns on your original investment but also on gains that accumulate over time.

The concept of compounding finds its application across a diverse array of financial instruments, including savings accounts, certificates of deposit (CDs), bonds, mutual funds, and retirement accounts. By diligently reinvesting the earnings, both individuals and investors can potentially build a sustained, long-term financial growth and the gradual accumulation of wealth. This time-tested mechanism exemplifies the profound impact that compounding can have on fostering financial prosperity.

3.1.1 The Basic Concept of Compounding

The key element in compounding is time. As the investment grows and generates returns, those returns are reinvested, leading to a larger investment base for the next period. Over time, this compounding effect can result in exponential growth.

Mathematically, the basic compounding formula is given by:

$$V_t = V_0 \times (1 + r)^t, \tag{3.1}$$

where V_t is the total value at the end of the tth time period, V_0 is the initial amount invested or deposited in the financial institution at the beginning of the period, r is the rate of interest expressed as an interest rate per time period, and t is the number of interest-bearing time periods, usually expressed in years.

It can be easily recognized that Equation (3.1) is an exponential nonlinear function of the form $y = ab^x$, where in this case $y = V_t$, $a = V_0$, $1 + r = b$ and $t = x$. If we take log-transformation on equation (3.1), we can obtain a linear model as follows:

$$log(V_t) = log(V_0) + t \times log(1 + r), \tag{3.2}$$

We can further re-write equation (3.2) as

$$y_t = \beta_0 + \beta_1 \times t, \tag{3.3}$$

Where $y_t = log(V_t)$, $\beta_0 = log(V_0)$ and $\beta_1 = log(1 + r)$. Equation (3.3) is now a simple linear regression described in Chapter 2 where the response variable is the log-transformed total value and the independent variable is the number of time period t. The intercept β_0 is the log-transformed initial deposited value and the slope parameter β_1 captures the information about the appreciation rate. Typically, these parameters should be estimated from financial data.

TABLE 3.1
Statement of Investment Growth with Annually Compounded Interest.

t	Vt
1	108145
2	116624
3	125453
4	136133
5	146638

3.1.2 Data Available

As a typical example in financial compounding, we use Mrs. Smith's data for the purpose of illustration. Mrs. Smith had an inheritance which was invested by her parents long time ago in a financial annuity with varying interest rates at Morgan Stanley. She has received the financial statement every year showing how the investment (in US dollars) has been growing at the end of every year for five years, which is created in the *R* dataframe *CompData* as follows:

```
# The available data
CompData = data.frame(t  = c(1,2,3,4,5),
  Vt = c(108145.25,116623.51,125453.43,136132.63,146638.48))
# Print the data
CompData
```

```
##   t     Vt
## 1 1 108145
## 2 2 116624
## 3 3 125453
## 4 4 136133
## 5 5 146638
```

This dataframe can be further displayed in table format as shown in Table 3.1 and also be displayed graphically in Figure 3.1 to show the compounding growth.

As seen in Table 3.1 and Figure 3.1, we can see the total amount earned (V_t) at the end of each year (i.e., $t = 1, 2, 3, 4, 5$), which is mostly characterized by the nonlinear exponential growth model described in Equation (3.1). The above amounts include both the principals and interest earned even we do not know the initial amount invested (i.e., V_0) and annualized interest rate (i.e., r), which are to be estimated from this data.

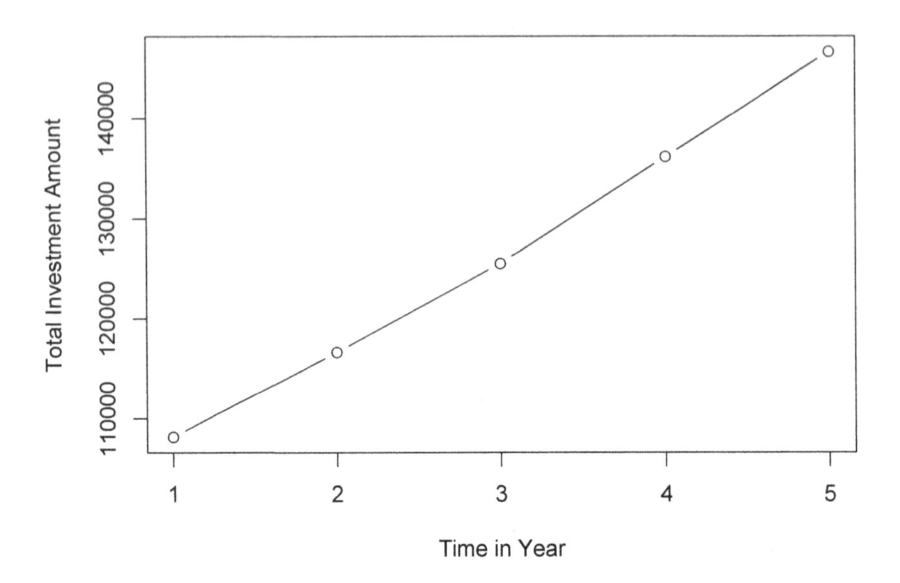

FIGURE 3.1
Relationship between the Investment Growth and Time

The growth of the invested amount through compounding follows a characteristic nonlinear exponential function over time. As we discuss this concept with Mrs. Smith, several pertinent questions arise in the discussion:

1. **Average Yearly Interest Rate from Morgan Stanley** (r): A crucial inquiry would revolve around the long-term average interest rate per year (denoted as r) obtained from Morgan Stanley. Understanding this interest rate is essential to gauging the pace at which her investment will grow over the years.

2. **Initial Investment Amount from Parents** (V_0): It's important to inquire about the initial amount (represented as V_0) that Mrs. Smith's parents invested initially. This initial investment forms the bedrock upon which the compounding process is built.

3. **Prediction and Estimation of Future Amount** (V_{10}): An intriguing financial question emerges regarding the projection and estimation of the total amount Mrs. Smith could get at the culmination of the tenth year (denoted as V_{10}). By factoring in the average yearly interest rate and the initial investment, we can employ the principles of compounding to foresee the growth trajectory of her investment over the specified time frame. This forecast yields valuable insights into the potential financial outcome of her future.

3.1.3 Computation and *R* Implementation

To answer the three questions above, we can use the linear regression estimation to estimate the parameters β_0 and β_1 associated in equation (3.3) to estimate the initial investment amount V_0 and the annual interest rate r.

This can be done using the *R* function *lm* as follows with equation (3.3):

```
# Using regression "lm"
estMod = lm(log(Vt)~t, CompData)
# Print the summary of model fitting
summary(estMod)
```

```
##
## Call:
## lm(formula = log(Vt) ~ t, data = CompData)
##
## Residuals:
##         1         2         3         4         5
##  1.02e-03  1.26e-04 -3.26e-03  2.07e-03  4.44e-05
##
## Coefficients:
##             Estimate Std. Error t value Pr(>|t|)
## (Intercept) 1.15e+01   2.42e-03    4761  2.0e-11 ***
## t           7.64e-02   7.29e-04     105  1.9e-06 ***
## ---
## Signif. codes:
## 0 '***' 0.001 '**' 0.01 '*' 0.05 '.' 0.1 ' ' 1
##
## Residual standard error: 0.00231 on 3 degrees of freedom
## Multiple R-squared:      1,    Adjusted R-squared:      1
## F-statistic: 1.1e+04 on 1 and 3 DF,  p-value: 1.92e-06
```

As seen from the model fitting summary, the regression model is statistically highly significant with p-value < 0.0001 and $R^2 = 0.999$. Using the model fitting, we can answer the three questions for Mrs. Smith.

As indicated by the summary of model fitting, the regression model demonstrates a remarkable level of statistical significance, as evidenced by a p-value that is notably less than 0.0001. Additionally, the coefficient of determination (R^2) stands at an impressive value of 0.999. These findings collectively signify the satisfactory fit of the regression model.

With this fitted model, we can now provide accurate answers to the three pertinent questions that pertain to Mrs. Smith's financial scenario.

First, let's extract the estimated parameters of β_0 and β_1 as follows:

```
# Get the estimated coefficients
est.betas = as.numeric(coef(estMod))
# Get beta0
beta0 = est.betas[1]; beta0
```

```
## [1] 11.51
```

```
# Get beta1
beta1 = est.betas[2]; beta1
```

```
## [1] 0.07637
```

Now we can answer the questions respectively as follows:

1. The long-term average interest rate per year (i.e., r) from Morgan Stanley, can be estimated as follows:

   ```
   # Estimate interest rate
   r = exp(beta1)-1;r
   ```

   ```
   ## [1] 0.07936
   ```

   As seen from this estimation, the estimated annualized interest rate is about 8%.

2. The initial amount (i.e., V_0) invested at the beginning from her parents can be estimated as follows:

   ```
   # Estimate initial investment
   V0 = exp(beta0);V0
   ```

   ```
   ## [1] 100092
   ```

   Therefore, the initial investment amount is about $100,092.

3. To predict and estimate the total amount Mrs. Smith will have at the end of the tenth year (i.e., V_{10}), we can use the estimated regression model as follows:

```
# Amount at the 10th year
V10 = exp(beta0+beta1*10);V10

## [1] 214811
```

This is to say that with the initial investment of $100,092 at the interest rate of 8%, the estimated total amount is at $214,811.4, which would be more than double the initial investment of $100,092. This is the principle of compounding.

3.2 Depreciation of Non-Current Assets

3.2.1 The Reduced Balance Method

Francis (2004) described the reducing balance method that if the original book value B is subjected to reducing balance depreciation at the rate $100r\%$ over T period, the depreciated value at the end of the T–th time period is given by:

$$DV = B \times (1 - r)^T, \tag{3.4}$$

where DV is the depreciated value at the end of the T-th time period, B is the original book value (original purchase price of the asset), r is the depreciation rate as a percentage, and T is the number of time period normally years.

It is easy to recognize that equation (3.4) is an exponential nonlinear function of the form $y = ab^x$, where in this case $y = DV$, $a = B$, $1 - r = b$ and $T = x$. If we take log-transformation on equation (3.4), we can obtain a linear model as follows:

$$log(DV) = log(B) + T \times log(1 - r), \tag{3.5}$$

We can further re-write equation (3.5) as

$$log(DV) = \beta_0 + \beta_1 \times T, \tag{3.6}$$

where $\beta_0 = log(B)$ and $\beta_1 = log(1-r)$. Equation (3.6) is now transformed into a simple linear regression described in Chapter 2 where the response variable is the log-transformed depreciated value and the independent variable is the number of time period T. The intercept β_0 is the log-transformed original book value of the assets and the slope parameter β_1 captures the information about the depreciation rate. Typically, these parameters should be estimated from financial data.

TABLE 3.2
Depreciation of Non-Current Assets (in US Dollar)

T	DV
1	12610
2	9938
3	7879
4	6533
5	5242
6	4134

3.2.2 Data Available

The following data was extracted by an auditor from an asset registered to JC Media limited. The asset was a certain non-current asset that was bought six years ago and the company utilized the reducing balance methods to charge depreciation. This data can be inputted to *R* to create an *R* dataframe *DepData* as follows:

```
# The available data
DepData = data.frame(T = c(1,2,3,4,5,6),
DV= c(12610.4, 9938.3, 7878.8, 6532.9,  5242.3, 4133.8))
# Print the data
DepData
```

```
##   T    DV
## 1 1 12610
## 2 2  9938
## 3 3  7879
## 4 4  6533
## 5 5  5242
## 6 6  4134
```

This dataframe can be further displayed in table format as seen in Table 3.2 and can also be displayed graphically in Figure 3.2 to show the depreciation over time.

As seen in Table 3.2 and Figure 3.2, the depreciation amount (DV) at the end of each year (T) is a nonlinear exponential decay function of time.

Also we do not know both the depreciation rate (r) and the initial book value in year 0 (B). But, we can use the linear regression model to estimate these parameters β_0 and β_1 associated in equation (3.6) to estimate the initial book value B and the depreciation rate r. This can be done using the *R*

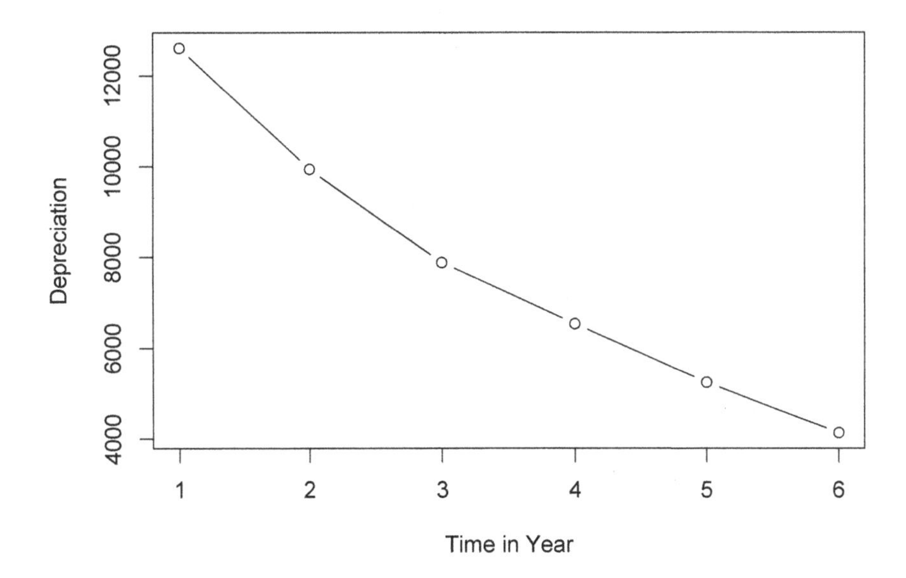

FIGURE 3.2
Depreciation over Time

function *lm* as follows with equation (3.6) to estimate an approximate linear relationship of the above exponential decay in net book value in this accounting problem associated with depreciation of noncurrent assets. With the estimated parameters, the auditor can then use them to verify some of the accounting questions as follows:

1. The depreciation rate (r) that the JC Media Limited used,
2. The initial book value (B) of this asset,
3. The expected depreciated value (net book value) by the end of the tenth year (BV_{10}),
4. The accumulated depreciation by the end of the tenth year.

3.2.3 Computation and *R* Implementation

To answer above questions, we can use the linear regression estimation to estimate the parameters β_0 and β_1 associated in equation (3.6) to estimate the initial book value B and the depreciation rate r.

This can be done using the *R* function *lm* as follows with equation (3.6):

```
# Using regression "lm"
estMod2 = lm(log(DV)~T, DepData)
```

```
# Print the summary of model fitting
summary(estMod2)
```

```
##
## Call:
## lm(formula = log(DV) ~ T, data = DepData)
##
## Residuals:
##        1        2        3        4        5        6
##  0.01110 -0.00752 -0.02023  0.01196  0.01138 -0.00668
##
## Coefficients:
##             Estimate Std. Error t value Pr(>|t|)
## (Intercept)   9.6507     0.0140     689  2.7e-11 ***
## T            -0.2195     0.0036     -61  4.3e-07 ***
## ---
## Signif. codes:
## 0 '***' 0.001 '**' 0.01 '*' 0.05 '.' 0.1 ' ' 1
##
## Residual standard error: 0.015 on 4 degrees of freedom
## Multiple R-squared:  0.999,  Adjusted R-squared:  0.999
## F-statistic: 3.72e+03 on 1 and 4 DF,  p-value: 4.32e-07
```

As seen from the model fitting summary, the regression model is statistically highly significant with p-value < 0.0001 and $R^2 = 0.999$. Using the model fitting, we can help the accounting auditor to answer the three questions.

First, let's extract the estimated parameters of β_0 and β_1 as follows:

```
# Get the estimated coefficients
est.betas = as.numeric(coef(estMod2))
# Get beta0
beta0 = est.betas[1]; beta0
```

```
## [1] 9.651
```

```
# Get beta1
beta1 = est.betas[2]; beta1
```

```
## [1] -0.2195
```

Now we can answer the questions respectively:

1. The depreciation rate (r) that the JC Media Limited used can be estimated as follows:

   ```
   # Estimate depreciation rate
   r = 1-exp(beta1);r
   ```

   ```
   ## [1] 0.1971
   ```

   which is about 20%.

2. The initial book value (B) of this asset can be estimated as follows:

   ```
   # The initial book values estimated
   B = exp(beta0);B
   ```

   ```
   ## [1] 15533
   ```

   which is \$15,532.5.

3. The expected depreciated value (net book value) by the end of the tenth year (BV_{10})

   ```
   # Estimate expected depreciated value
   DV10 = exp(beta0+beta1*10);DV10
   ```

   ```
   ## [1] 1730
   ```

   Therefore, the depreciated value by the end of the tenth year is at \$1,729.5.

4. The accumulated depreciation by the end of the tenth year can be calculated by the difference between the initial book value (B) when the asset was first acquired and the net book value at the end year ten (DV_{10}). Therefore accumulated depreciation at the end of the tenth year is $B - DV_{10} = 15{,}532.5 - 1729.5 = \$13{,}803$.

3.3 Pros and Cons of Linearization

Linearizing by transforming exponential nonlinear models into linear models offers both advantages and disadvantages, each carrying implications for the accuracy and interpretability of the financial data analysis. Some of the *pros* and *cons* can be summarized as follows. The associated *pros* are:

1. *Simplicity and Interpretability*: Linear models are generally easier to understand and interpret compared to complex nonlinear equations. Transforming an exponential relationship into a linear model can simplify the model's structure, making it more accessible to financial practitioners and managers.
2. *Compatibility with Linear Regression*: Linear regression is a well-established statistical technique with widely available software and resources as discussed in Chapter 2. Converting to a linear model allows the utilization of familiar linear regression tools for parameter estimation and hypothesis testing.
3. *Interpretation of Coefficients*: In linear models, coefficients have straightforward interpretations. A unit change in the predictor variable corresponds to a constant change in the response variable, making it easier to derive actionable insights from the results.

The associated *cons* can be:

1. *Information Loss*: Converting a complex exponential financial relationship into a linear one might entail information loss. In general, nonlinear relationships often capture nuances that linear transformations and approximations may overlook, leading to potential inaccuracies in predictions and explanations.
2. *Validity Assumption*: Transforming data to fit a linear model assumes that the underlying relationship is truly linear. If the true relationship is genuinely nonlinear, forcing a linear approximation can introduce bias and misleading conclusions.
3. *Extrapolation Risk*: Linear models may not accurately capture the behavior of the original nonlinear exponential data beyond the range of observed values. Extrapolating predictions to extreme values can result in unreliable forecasts and erroneous decision-making.
4. *Residual Patterns*: Transforming data can lead to unusual patterns in residuals (the differences between observed and predicted values),

violating the assumptions of linear regression. This can impact the validity of statistical inferences and compromise the model's quality.

Generally speaking, transforming exponential nonlinear financial models into linear forms presents a trade-off between simplicity and fidelity to the underlying data. While linear models offer ease of interpretation and compatibility with established techniques, they may sacrifice accuracy and introduce potential biases. The decision to make such a conversion should be guided by a thorough understanding of the data, the goals of analysis, and a critical evaluation of the *pros* and *cons* outlined above.

Close investigations should be made to the above *pros* and *cons* and we will investigate the residual patterns from both models using a Monte-Carlo simulation study. Interested readers can use the logic and the *R* code developed in this Chapter to investigate the other *pros* and *cons* listed above.

3.3.1 Theoretical Differences in Residual Patterns

To initiate the investigation, let's first examine the *Financial Compounding Model* in Section 3.1 and the model on *Depreciation of Non-Current Assets* in Section 3.2. It can be easily recognized that both models can be expressed as an exponential nonlinear function of the form

$$y = a \times b^x$$

where (a, b) are the parameters. If we take log-transformation on this exponential model, we obtain the associated linear model as follows:

$$log(y) = log(a) + log(b) \times x.$$

With observed data, we perform linear regression to model $log(y) = log(a) + log(b) \times x$ with least squares estimation, which is essentially assumed that we fit a model

$$\begin{aligned} log(y) &= log(a) + log(b) \times x + \epsilon_L \\ &\equiv \beta_0 + \beta_1 \times x + \epsilon_L \end{aligned} \tag{3.7}$$

where $\beta_0 = log(a)$ and $\beta_1 = log(b)$ are the regression parameters. The residual term in this linearized model ϵ_L is assumed to be normally distributed with mean 0 and variance of σ_L^2. Transforming the linear model in Equation (3.7) back to the original values, we in fact have a model as follows:

$$y = a \times b^x \times e^{\epsilon_L} \equiv a \times b^x \times \epsilon. \tag{3.8}$$

In this model (3.8), the error term $e^{\epsilon_L} \equiv \epsilon$ is multiplicative to the nonlinear exponential model $y = a \times b^x$. This multiplicative error term is now *log-normally* (**not normally** anymore) distributed with mean 0 and variance of σ_L^2.

In contrast, if the original nonlinear exponential model is fitted to the data, the underlying model is in fact as follows:

$$y = a \times b^x + \epsilon_N \tag{3.9}$$

where error term ϵ_N in model (3.9) is additive to the nonlinear exponential model $y = a \times b^x$, which is assumed to be normally distributed with mean 0 and variance of σ_N^2.

This is the case of *Residual Patterns* difference, where there's a fundamentally theoretical difference. These two models differ significantly in terms of their underlying assumptions about the distribution of residuals. The original nonlinear exponential model operates with an additive error structure. In contrast, when transforming it into a linear model, the error structure becomes *multiplicative.* This divergence in assumptions about how errors are distributed can lead to notable differences in the results of these models.

We will numerically demonstrate these differences with a Monte-Carlo simulation study.

3.3.2 Numerical Differences in Residual Patterns

3.3.2.1 Simulation Setup

To show the numerical difference, we will generate data from the original nonlinear exponential model in equation (3.9) with additive error structure where the true parameters (a, b, σ_N) are equal to (2,2,2).

We then fit the generated data with two estimation approaches:

1. *Linear Regression* with *R* function *lm* to the transformed linear model in Equation (3.7), or equivalently in equation (3.8), as what we did in the previous sections,
2. *Nonlinear regression* with *R* function *nls* (i.e., *nonlinear least squares* estimation) to fit the model in equation (3.9), which leads to nonlinear regression to be discussed in Chapter 4.

3.3.2.2 Just One Simulation

Let's start with one simulation to show the simulated data and the associated model fitting. The following *R* code chunk can be used to simulate the data:

```
# Set random seed for reproducibility
set.seed(3388)
# The true parameters
a = b = 2; sigmaN = 2
```

```
# Let's generate observations with n sample
n = 100
# Let's generate x from standard normal and sort it
x = sort(rnorm(n, mean=3));
# The  we can generate the additive error term
errN = rnorm(n, mean=0, sd = sigmaN);
# Then we can get the outcome y
y = a*b^x +errN;
# Make a dataframe and print the first 6 observations
Simu.data=data.frame(x=x,error=errN,true.y=a*b^x,obs.y=y)
head(Simu.data)
```

```
##        x   error true.y obs.y
## 1 0.9069  1.5916  3.750 5.341
## 2 1.1433  0.7051  4.418 5.123
## 3 1.2655 -3.4248  4.808 1.383
## 4 1.2927 -3.1275  4.900 1.772
## 5 1.3292  1.9934  5.025 7.019
## 6 1.3878  0.4116  5.233 5.645
```

With the generated data above, we can fit the nonlinear exponential model with *nls* and the associated transformed linear model with *lm* as follows:

```
# Fit the Nonlinear model using nls
fitNLM = nls(y~a*b^x, start=list(a=1, b=1), trace=T)
```

```
## 4.968e+04 (4.31e+00): par = (1 1)
## 4.886e+04 (9.26e+00): par = (0.3405 1.413)
## 4.064e+04 (9.92e+00): par = (0.2436 2.024)
## 3.134e+04 (8.70e+00): par = (0.4538 2.017)
## 1.782e+04 (6.53e+00): par = (0.8219 2.016)
## 4760.     (3.26e+00): par = (1.374 2.016)
## 408.6     (1.94e-05): par = (1.926 2.016)
## 408.6     (3.50e-08): par = (1.926 2.016)
```

```
# Print the model fit
summary(fitNLM)
```

```
##
## Formula: y ~ a * b^x
##
## Parameters:
##   Estimate Std. Error t value Pr(>|t|)
## a   1.9262     0.0935    20.6   <2e-16 ***
```

```
## b   2.0165     0.0238    84.8   <2e-16 ***
## ---
## Signif. codes:
## 0 '***' 0.001 '**' 0.01 '*' 0.05 '.' 0.1 ' ' 1
##
## Residual standard error: 2.04 on 98 degrees of freedom
##
## Number of iterations to convergence: 7
## Achieved convergence tolerance: 3.5e-08
```

```
# Fit the transformed linear model with log-transformation
fitLM = lm(log(y)~x)
# Print the model fit
summary(fitLM)
```

```
##
## Call:
## lm(formula = log(y) ~ x)
##
## Residuals:
##     Min      1Q  Median      3Q     Max
## -1.1604 -0.0631 -0.0058  0.1220  0.5033
##
## Coefficients:
##             Estimate Std. Error t value Pr(>|t|)
## (Intercept)   0.5675     0.0830    6.84  6.9e-10 ***
## x             0.7250     0.0268   27.05  < 2e-16 ***
## ---
## Signif. codes:
## 0 '***' 0.001 '**' 0.01 '*' 0.05 '.' 0.1 ' ' 1
##
## Residual standard error: 0.247 on 98 degrees of freedom
## Multiple R-squared:  0.882,  Adjusted R-squared:  0.881
## F-statistic:  732 on 1 and 98 DF,  p-value: <2e-16
```

```
# Transform back for the estimated parameters
estLM.par = exp(coef(fitLM))
# Now get the predicted values for both models
pred.NLM = predict(fitNLM,x)
pred.LM  = exp(predict(fitLM))
# Plot the fittings and overlay the fits
# 1. Plot the data in Figure 3.3
plot(y~x, las=1, xlab="The Independent Variable (x)",
     ylab="The Exponential Response Variable (y)")
```

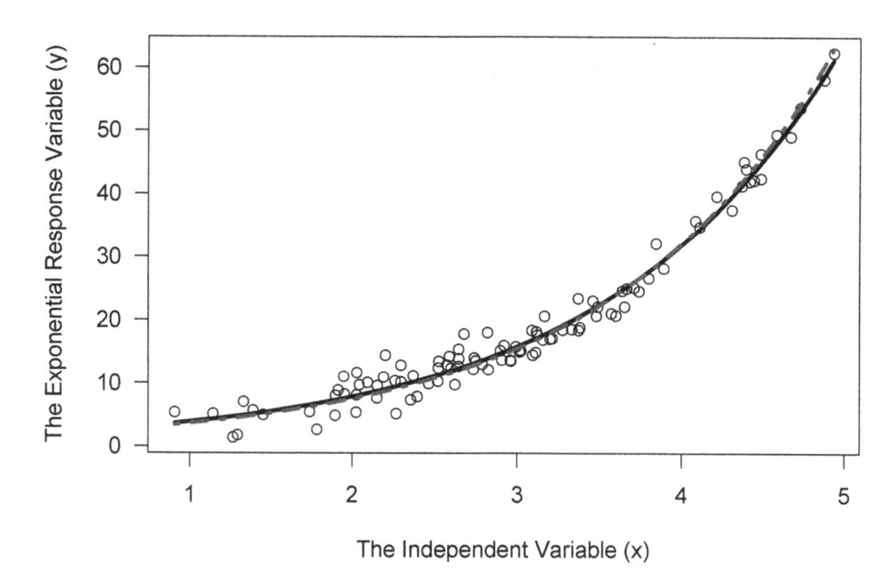

FIGURE 3.3
Illustration of Data Generated from Nonlinear Exponential Model and the Two Fitted Models

```
# Overlay the nonlinear model fitting with the "black" line
lines(x,pred.NLM, lty=1, lwd =3, col="black")
# Overlay the linear model fitting with "red" dashed line
lines(x,pred.LM,  lty=4, lwd =3,col="red")
```

As seen from the nonlinear regression fitted by *nls*, the estimated parameters for (a, b) are $(\hat{a}, \hat{b}) = (1.92623, 2.01648)$, which are statistically highly significant with p-values < 0.0001. The estimated residual standard error is $\hat{\sigma}_N = 2.042$ with 98 degrees of freedom since we have $n = 100$ observation to estimated 2 parameters of a and b. Note that those estimates are very close to the true values of $(a, b, \sigma_N) = (2, 2, 2)$, which is supposed to occur since we generated data from this additive model.

With the linearized model in (3.7), the estimated regression parameters are $\left(\hat{\beta}_0, \hat{\beta}_1\right) = (0.56749.0.72497)$. This can be transformed to obtain $\left(\hat{a}, \hat{b}\right) = (1.763834, 2.064667)$, which is different from the true values of $(a, b) = (2, 2)$. (Note that the estimated σ has a very complicated formula and we will omit it in this comparison).

Figure 3.3 graphically illustrated the data generation and overlaid by the two model fittings, where the nonlinear model fitting is overlaid by the *black* solid line and the linear model fitting is overlaid by the dashed *red* line. It seems that both models fitted the data well, but due to the scales of the data, the linearized model (as shown in *red* dashed line) exhibits lack of fit, which evidenced by the $\left(\hat{a}, \hat{b}\right) = (1.763834, 2.064667)$, which is different from the true values of $(a, b) = (2, 2)$.

3.3.2.3 Full-Scale Monte-Carlo Simulations

To see the systematic biases in parameter estimation, we can conduct a full-scale Monte-Carlo simulation with a large number of simulations (say 10,000). This can be implemented easily in *R* as follows:

```
# Set random seed for reproducibility
set.seed(3388)
# The true parameters
a = b = 2; sigmaN = 2
# Let's generate observations with n sample
n = 100
# The number of simulations
nsimu=10000
# Create a temporal matrix to hold the simulated parameters
estParMat = matrix(0, nrow=nsimu, ncol=4)
colnames(estParMat) = c("aN","bN","aL","bL")
# Now loop over the simulation
for(s in 1:nsimu){
# Let's generate x from standard normal and sort it
x = sort(rnorm(n, mean=3));
# The  we can generate the additive error term
errN = rnorm(n, mean=0, sd = sigmaN);
# Then we can get the outcome y
y = a*b^x +errN;
# Fit the Nonlinear model using nls
fitNLM = nls(y~a*b^x, start=list(a=2, b=2))
# Output the estimated parameters
estParMat[s,1:2] = coef(fitNLM)
# Fit the transformed linear model with log-transformation
fitLM = lm(log(y)~x)
# Transform back for the estimated parameters
estLM.par = exp(coef(fitLM))
estParMat[s,3:4] = estLM.par
} # end of s-loop
```

```
# Print the first 6 simulations
head(estParMat)
```

```
##          aN    bN    aL    bL
## [1,] 1.926 2.016 1.764 2.065
## [2,] 2.021 1.986 1.749 2.071
## [3,] 1.955 2.010 1.657 2.102
## [4,] 1.876 2.035 1.439 2.190
## [5,] 2.014 1.998 1.986 1.998
## [6,] 2.056 1.988 1.788 2.067
```

With these 10,000 simulations, we can calculate the sampling means from these 10,000 samples using *R* function *apply* as follows:

```
apply(estParMat,2,mean)
```

```
##    aN    bN    aL    bL
## 2.001 2.000 1.852 2.046
```

This shows that the overall mean for the estimated $(\hat{a}, \hat{b}) = (2.000863, 2.000177)$ for the original nonlinear model estimated with *nls* function and (1.852018, 2.046139) for the log-transformed linear model estimated with *lm* function, respectively, where the true parameters are $(a, b) = (2{,}2)$. This shows that the estimated regression parameters for the original nonlinear model estimated with *nls* are unbiased, but they are biased with the linearized model estimated using linear regression *lm*. This evidence can be graphically shown in the following Figure 3.4.

3.3.3 Additive Errors and Multiplicative Errors

The above investigation leads to a common question in financial modeling on which error distributions should be used. Should we choose additive residual errors or multiplicative residual errors?

In financial modeling, both additive and multiplicative errors are used to describe how the actual values of financial outcomes differ from the predicted or modeled values. These concepts are essential in understanding the accuracy and reliability of financial models.

The *additive errors* refer to discrepancies between the predicted value and the actual value that are constant and independent of the predicted value itself. In other words, the error remains the same regardless of the magnitude of the prediction. *Additive errors* are often encountered when there are fixed external factors influencing the observed values. In financial modeling, they can occur when there are consistent factors affecting outcomes that are not accounted

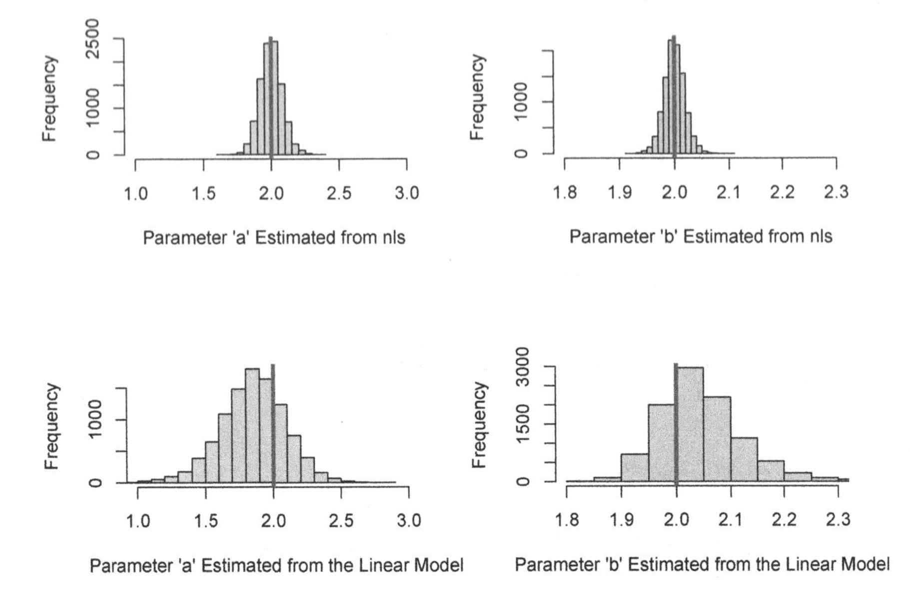

FIGURE 3.4
Monte-Carlo Distributions Overlaid with the True Parameters

for in the model. Addressing additive errors might involve adjusting the model or incorporating additional variables to better capture these influences. *Multiplicative errors*, on the other hand, occur when the discrepancy between the predicted value and the actual value is proportional to the predicted value itself. As the predicted value increases, the error also increases proportionally. *Multiplicative errors* can stem from situations where percentage discrepancies or proportional differences are more meaningful than absolute differences. In financial modeling, they might arise when dealing with variables that exhibit exponential growth or where relative changes matter more than absolute deviations.

Understanding whether errors are additive or multiplicative is crucial for accurately interpreting the performance of financial models. Different modeling techniques and strategies might be needed to address each type of error so correct model can be chosen to improve the model's predictive power. Additionally, the choice between additive and multiplicative modeling can have significant implications for risk assessment, investment decisions, and other financial analyses.

Additive errors and *multiplicative errors* in financial modeling are relevant in different scenarios, depending on the nature of the data and the specific context of the financial analysis. Here we provide a few general recommendations to considered.

Consider *Additive Errors* if :

1. *Stable External Factors*: When there are consistent external factors that influence the observed values but are not captured in the model, *additive errors* might be more appropriate. These factors contribute a fixed amount to the error regardless of the level of the predicted value.
2. *Small-Scale Variations*: *Additive errors* are suitable when the variations between predicted and actual values are relatively small and do not significantly change with changes in the predicted values.
3. *Interpreting Residuals*: In linear regression and other modeling techniques, if the residuals (differences between actual and predicted values) show a consistent, constant pattern across different levels of predictions, it indicates the presence of *additive errors.*

Consider *Multiplicative Errors* if:

1. *Proportional Deviations*: When the discrepancies between predicted and actual values are proportional to the predicted values, multiplicative errors are more appropriate. This occurs when the relative difference between predicted and actual values is more meaningful than the absolute difference.
2. *Exponential Growth*: In cases where the underlying process exhibits exponential growth or compounding, such as in investment returns, stock prices, or interest rates, *multiplicative errors* might be a better fit.
3. *Percentage Changes*: When understanding the percentage change or relative change is more important than absolute change, as is often the case in financial analyses, *multiplicative errors* are preferred.
4. *Logarithmic Transformations*: In some cases, transforming the data using logarithms can help convert multiplicative relationships into additive ones, making them amenable to linear modeling.

It's important to note that the choice between additive and multiplicative errors is not always straightforward and may require careful examination of the data and domain expertise in finance. Additionally, different modeling techniques can be used to account for each type of error. In practice, understanding the nature of errors is crucial for accurate predictions and meaningful interpretations in financial modeling.

3.4 Discussions

This chapter serves as a transition from linear models to nonlinear models by transforming nonlinear financial models to their linear counterparts. With the appropriate transformation, familiar linear regression techniques can be used to solve complicated nonlinear models.

For example, some exponential nonlinear models can be transformed into their linear forms. These can then be easily implemented into linear regression analysis to estimate associated key parameters and interpret parameters in financial decision-making. We focused on two of the most commonly used nonlinear models including the depreciation of non-current assets using the reducing balance method and financial compounding.

The approaches illustrated in this chapter can be generalized to most nonlinear models in accounting and finance that are in exponential form. However, there are many financial models which can not be simply transformed into linear models and nonlinear modeling techniques are needed and this is the topic in next Chapter 4.

3.5 Exercises

1. Refit the data in Section 3.1 with nonlinear regression. Hint: use *R* function *nls*.
2. Refit the data in Section 3.2 with nonlinear regression. Hint: use *R* function *nls*.
3. Follow the steps in Section 3.3.2.1, but simulate data from the log-linear model in equation (3.7) with additive error structure where the true parameters (a, b, σ_L) are equal to (1,1,1) to investigate the effect of the nonlinear model estimation with equation (3.9).

4

Nonlinear Regression Modeling

In the previous Chapter 3, we explored how to transform nonlinear models into linear models in financial data, especially when they exhibit exponential forms. While linear regression and transformations like logarithms can handle many scenarios, certain financial models are inherently nonlinear due to their intricate dynamics. These models might involve interactions between variables, non-additive effects, or non-constant relationships over time. For example in option pricing, nonlinear models are essential for accurately pricing options and derivatives. Stochastic volatility models, for instance, capture the changing volatility dynamics inherent in financial markets. Similar to credit risk assessment, nonlinear techniques help assess credit risk by modeling the complex interactions between various factors influencing creditworthiness. In portfolio optimization, portfolio selection often involves nonlinear relationships between assets' expected returns and risks. Nonlinear models aid in finding optimal portfolios that balance these factors. Further in behavioral finance, understanding investor behavior often requires nonlinear models to capture biases, heuristics, and sentiment-driven decision-making.

Attempting to force these commonly used financial models into linear models may lead to misleading results. In reality, there are many more instances where financial models defy simple transformations to linear models. In such cases, advanced nonlinear modeling techniques become indispensable.

This chapter delves into these techniques, highlighting their applications in accounting and finance. Since there are so many nonlinear models in finance which we cannot cover in one chapter, we are going to focus on the logistic growth model with applications to model the adoption and saturation of financial products, services, and market trends. Interested readers can refer to Bates and Watts (1998) for other types of nonlinear models and the associated fitting methods.

4.1 Logistic Growth Model

The logistic growth model is a type of mathematical model that can be used to describe population dynamics. Such populations can be financial markets,

DOI: 10.1201/9781003469704-4

human populations, disease populations, or any ecological and environmental populations.

Originally derived from biology, the logistic growth models have found meaningful applications in finance when describing adoption patterns, market penetration, and saturation of financial products or trends.

Let's denote $A(t)$ as the quantity of a financial product/portfolio at time t, this logistic growth model can be formalized by the differential equation:

$$\frac{dA(t)}{dt} = r \times A(t) \times \left(1 - \frac{A(t)}{C}\right), \tag{4.1}$$

where r is the growth rate parameter that determines how quickly the adoption occurs and C is the carrying capacity or the maximum value that $A(t)$ can reach.

We can solve this differential equation at (4.1) to obtain the solution as follows:

$$A(t) = \frac{CA_0 e^{rt}}{C + A_0\left(e^{rt} - 1\right)} = \frac{C}{1 + \left(\frac{C-A_0}{A_0}\right)e^{-rt}}, \tag{4.2}$$

where A_0 is the initial quantity of the financial product at time 0. It can be easily seen that $\lim_{t\to\infty} A(t) = C$, which is to say that C is the limiting value of the quantity of the financial product $A(t)$ as t approaches to infinity, i.e., the highest value that financial product can reach given infinite time (or close to reaching in finite time).

Denoting $t_0 = \frac{1}{r} ln\left(\frac{C-A_0}{A_0}\right)$, the logistic growth model in (4.2) can be reformulated as follows:

$$A(t) = \frac{C}{1 + e^{-r(t-t_0)}}. \tag{4.3}$$

This model in (4.3) is the so-called three-parameter (C, t_0, r) logistic growth curve model which is commonly used in biology and adopted in finance where C is the parameter of maximum capacity, r is the growth rate and t_0 is the change point or the tipping point at which growth starts to change from fast growth before t_0 to slowing growth afterward.

To use this model to analyze real financial data, an error term is added to the three-parameter logistic growth model (4.3) to be used for statistical analysis as follows:

$$A(t) = \frac{C}{1 + e^{-r(t-t_0)}} + \epsilon(t), \tag{4.4}$$

where $\epsilon(t)$ is assumed to be random and is normally distributed with a mean of 0 and a standard deviation of σ, i.e., $\epsilon(t) \sim N\left(0, \sigma^2\right)$.

4.2 Theory of Nonlinear Regression

4.2.1 R Packages for Nonlinear Regression

There are several *R* packages that provide tools for nonlinear regression and modeling. These packages are essential for implementing various nonlinear models and conducting complex data analysis in the field of finance and beyond. Here are a few notable *R* packages for nonlinear regression and modeling:

1. *Stats*: A collection of statistical functions by *R* Core Team and contributors worldwide.
2. *nls* Function (Base *R*): The *nls()* function in base *R* allows you to fit nonlinear regression models using the least squares method and *nls2* for nonlinear regression with brute force. They are the basic tools for a wide range of nonlinear modeling tasks.
3. *drc* (Dose-Response Curves): The *drc* package is primarily used for analyzing dose-response curves, which can have applications in areas like finance when studying the response of an entity to different levels of exposure or intervention.
4. *nlstools*: Tool sets for nonlinear regression analysis.
5. *nnls*: The Lawson-Hanson algorithm for non-negative least squares (NNLS).
6. *mgcv* (Generalized Additive Models for Location Scale and Shape): The *mgcv* package provides functionalities for fitting generalized additive models (GAMs), which can capture both linear and nonlinear relationships using smoothing functions. It's useful for analyzing complex relationships in financial data.
7. *nlme* (Linear and Nonlinear Mixed Effects Models): The *nlme* package specializes in linear and nonlinear mixed effects models, which are essential for analyzing hierarchical or clustered data with both fixed and random effects. It's relevant for longitudinal financial data and panel data analysis.
8. *nplr* (Nonparametric Low-Rank Estimation): The *nplr* package focuses on nonparametric modeling using low-rank estimation techniques. While not exclusively nonlinear regression, it can be used when the relationship between variables is complex and not well-defined.
9. *quantreg* (Quantile Regression): The *quantreg* package deals with quantile regression, a technique that goes beyond mean regression to

model the relationships at various quantiles of the response variable. This is useful for understanding different aspects of the relationship in financial data.

10. *keras* (Deep Learning): The *keras* package, although mainly designed for deep learning, can also handle complex nonlinear modeling tasks using neural networks. This can be relevant when dealing with large financial datasets.
11. *forecast* (Time Series Forecasting): The *forecast* package focuses on time series forecasting techniques. While not exclusively nonlinear, it can include models like exponential smoothing and state space models, which capture nonlinear time series behavior.

Note that the choice of *R* package and *R* function depends on the specific modeling requirements and data characteristics. Each package comes with its own set of features, functionalities, and learning curves. Interested readers can refer to Nash (2014) for details on other *R* tools for nonlinear modeling.

4.2.2 *R* Function *nls* for Nonlinear Regression

The *stats* package in *R* is a fundamental package that provides a wide range of statistical functionalities, including the *nls()* function for nonlinear regression. The *nls* is used to determine the nonlinear (weighted) least-squares estimates of the parameters of a nonlinear model (Bates and Watts, 1998). The general use of *nls* is as follows:

```
nls(formula, data, start, control, algorithm,
    trace, subset, weights, na.action, model,
    lower, upper, ...)
```

where

- *formula* is a nonlinear model formula including variables and parameters,
- *data* is an optional data frame in which to evaluate the variables in formula and weights,
- *start* is a list of starting estimates,
- *control* is an optional list of control settings for the optimization,
- *algorithm* is character string specifying the algorithm to use,
- *trace* is a logical value indicating if a trace of the iteration progress should be printed,
- *subset* is an optional vector specifying a subset of data to be used in the fitting process,

- *weights* is an optional numeric vector of known weights for the weighted least squares,
- *na.action* is a function on what should happen when the data contains NAs,
- *model* is a *TRUE/FALSE* logical on whether or not to return the model frame,
- *lower* and *upper* are vectors of lower and upper bounds.

After fitting the model using *nls()*, you can use the *summary()* function to obtain information about the fitted parameters, their estimated values, standard errors, t-values, and p-values. This summary output helps you assess the goodness of fit and the significance of the estimated parameters.

Keep in mind that fitting nonlinear models can be more challenging than linear models due to the potential for convergence issues and sensitivity to initial parameter values. Experimenting with different starting values, checking for convergence warnings, and understanding the domain context are essential parts of successfully using the *nls()* function for nonlinear regression.

Lastly, remember that the *nls()* function works well for relatively simple nonlinear models. For more complex models or situations with convergence issues, you might need to explore other nonlinear modeling packages or techniques, as listed above.

4.2.3 Theory of Nonlinear Regression in *nls*

For a general description of nonlinear regression, let's denote y_t, such as the quantity of financial product $A(t)$, at time $x(t)$) as the response and independent variables at their tth observation (where $t = 1, \cdots, n$). Then the general nonlinear model is described as follows:

$$y_t = E(y_t|x_t) = f(x_t, \theta) + \epsilon_t \tag{4.5}$$

In this model, the mean response of $E(y|x)$ depends on x through the mean function $f(x, \theta)$, where the predictor x can have one or more components and the parameter vector θ can also have one or more components. In the logistic growth model in (4.4), x consists of the single predictor *time* in t and the parameter vector $\theta = (C, r, t_0)$ has three components.

In the model (4.5), we further assume that the errors ϵ_t are independently distributed with error variance parameter σ^2/w_t, where σ^2 is an unknown, but w_i are assumed to be known as nonnegative *weights*. If $w_i = 1$, the weighted least squares regression simplifies the classical unweighted least squares regression.

In the *R* package *stats*, The *nls* function is to estimate the parameter vector θ that minimizes the residual sum of squares,

$$RSS(\theta) = \sum_{t=1}^{n} w_t \left[y_t - f(x_t, \theta)\right]^2. \tag{4.6}$$

We denote the estimated θ as $\hat{\theta}$. Unlike the straightforward least-squares estimation used in general linear regression as discussed in Chapter 2, nonlinear models often lack a closed-form analytical formula for calculating parameter estimates ($\hat{\theta}$) due to their inherent complexity. In such instances, iterative optimization procedures become necessary to derive these estimates from the nonlinear model $f(x, \theta)$. Consequently, the importance of having an appropriate initial value cannot be understated, as it significantly aids in the search for the optimal $\hat{\theta}$ estimate mentioned in equation (4.6).

By default, the *Gauss-Newton algorithm* is employed for this optimization process. However, alternative options are available to cater to different scenarios. The *plinear* option leverages the *Golub-Pereyra algorithm*, designed for partially linear least-squares models. Additionally, the *port* option taps into the *nl2sol* algorithm from the *Port* library, offering an alternative approach to achieving parameter estimates in these intricate nonlinear regression cases.

4.3 Nonlinear Regression for Mobile Banking

There are many applications in finance for logistic growth models. This chapter will focus on applications to model the adoption and saturation of financial products, services, or market trends, specifically for market penetration of financial products. For instance, consider a new mobile banking app (for example Zelle) introduced by a bank. Initially, the adoption might be slow, but as users recognize its benefits and word spreads, the adoption rate accelerates. Over time, as most potential users adopt the app, the growth levels off or stops.

4.3.1 Data on Customer Size on Mobile Banking

Data are available for a new mobile banking app (i.e., Zelle) introduced by a bank in a 3-year (i.e., 36 months) period with customers' size in unit of millions (denoted by *At*) and the growth rate (denoted by *Rate*). The dataset is saved in *mobileservice.csv* and we can load the data into the *R* session as follows:

```
# Read in the mobile banking data
dMobile = read.csv("data/mobileservice.csv", header=T)
```

```
# Print the data summary
summary(dMobile)
```

```
##       Month             At              Rate
##  Min.   : 1.00   Min.   :  1.8   Min.   :-0.970
##  1st Qu.: 9.75   1st Qu.: 85.3   1st Qu.:-0.024
##  Median :18.50   Median :417.6   Median : 0.032
##  Mean   :18.50   Mean   :361.2   Mean   : 0.951
##  3rd Qu.:27.25   3rd Qu.:636.3   3rd Qu.: 0.291
##  Max.   :36.00   Max.   :707.5   Max.   :20.643
##                                  NA's   :1
```

```
# Print the first 10 observations
head(dMobile,10)
```

```
##    Month      At     Rate
## 1      1  11.465       NA
## 2      2  18.025  0.57207
## 3      3  46.910  1.60253
## 4      4   5.903 -0.87417
## 5      5   6.474  0.09683
## 6      6  61.608  8.51558
## 7      7   1.832 -0.97026
## 8      8  39.654 20.64315
## 9      9  61.747  0.55716
## 10    10 112.497  0.82189
```

This dataset is to describe the quantity of customers (denoted by A_t) who are using the new mobile app for their banking, which can be graphically plotted as follow using the following R code chunk:

```
# Make the plot on population growth
par(mar = c(4, 4, .1, .1))
plot(At~Month, xlab="Month", ylab="Customer Size (in Millions)",
     data = dMobile,  pch = 19)
```

It can be seen from Figure 4.1 that the initial adoption was slow for the first half year, but as users recognize its benefits and word spreads, adoption accelerates. Over time, as most potential users adopt the app, growth levels off after 2 years. This is the typical nonlinear adoption trend in financial industries. A linear model would not be appropriate to describe this growth model, so a nonlinear regression model should be used, which is the typical pattern in the logistic growth models.

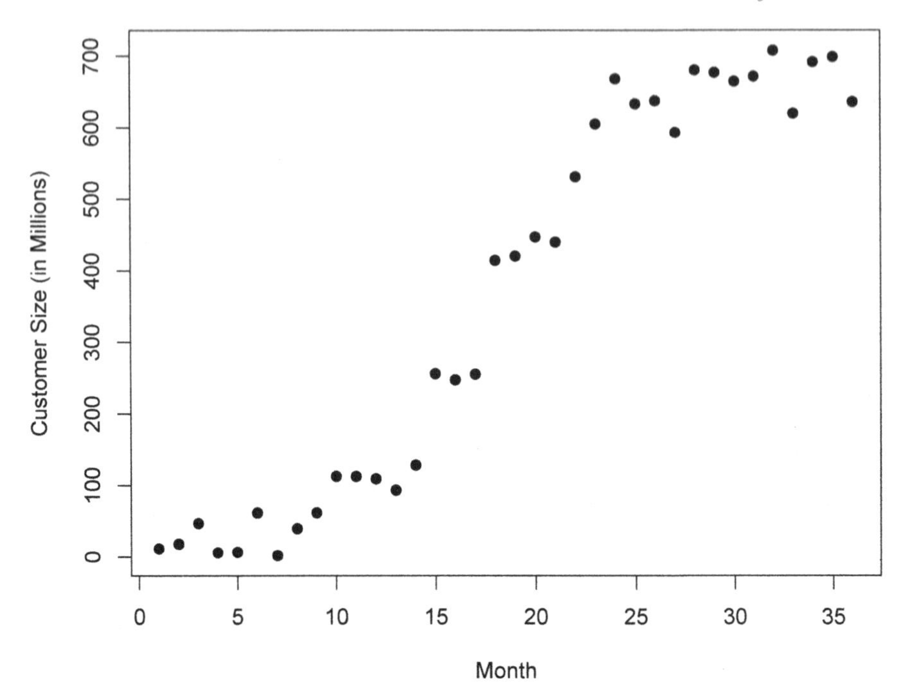

FIGURE 4.1
Customer Size in Mobile Banking

We can similarly illustrate the monthly growth rates with the following *R* code chunk:

```
# Plot the growth rates
par(mar = c(4, 4, .1, .1))
plot(Rate~Month, xlab="Month", ylab="Customer Growth Rate",
     data = dMobile, type = 'b', pch = 19)
# Add a horizontal line at zero growth
abline(h=0, col="red")
```

As seen from Figure 4.2, the initial growth rates were variably large during the first year or two, but level off to zero after 20 months.

4.3.2 Fitting the Mobile Banking Data Using *R* Function *nls*

Let's fit the data *dMobile* to the nonlinear logistic growth model in equation (4.3). Since we need a set of good initial values for *nls* to converge, we make an educated *guesstimate* that a set of initial values could be $C = 700$, $t_0 = 15$, $r = 0.3$ based on the Figure 4.1. That means that we assume that the parameter for the maximum carrying capacity of this mobile banking app can go to 500

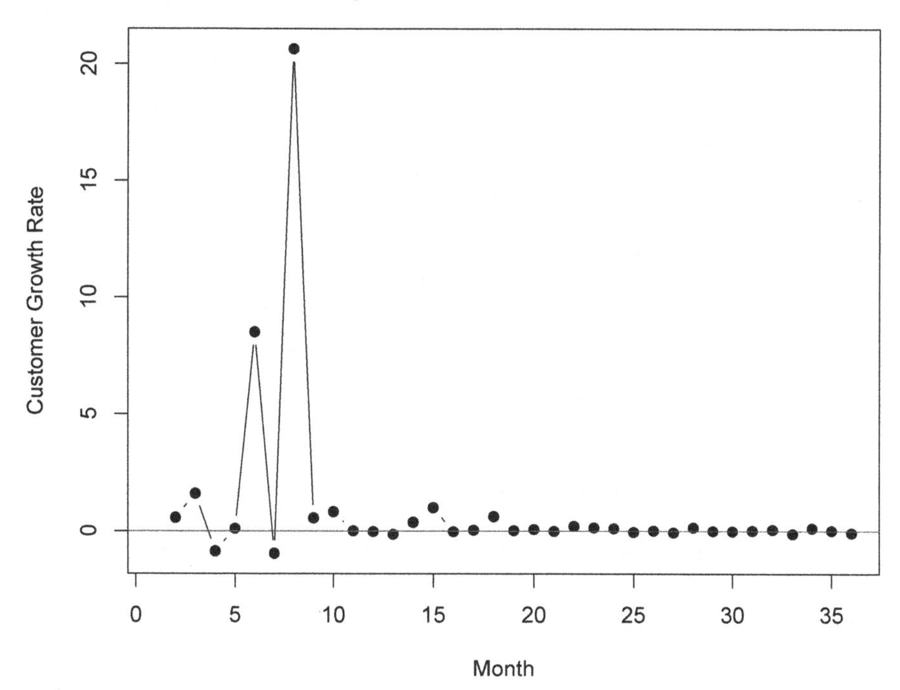

FIGURE 4.2
Mobile Banking Growth Pattern

million (i.e. $C = 700$), with a population growth rate of 30% (i.e. $r = 0.3$), and the initial change time point at about the month of 15 (i.e., $t_0 = 15$). With this set of parameters, we can run *nls* as follows:

```
# Use the default Gauss-Newton search algorithm
Mobile.mod = nls(At~C/(1+exp(-r*(Month-t0))),
                 data=dMobile, start=list(C=700, t0=15,r=0.3))
#Print the model fit
summary(Mobile.mod)
```

```
##
## Formula: At ~ C/(1 + exp(-r * (Month - t0)))
##
## Parameters:
##    Estimate Std. Error t value Pr(>|t|)
## C  684.0419    13.1085    52.2  < 2e-16 ***
## t0  17.6345     0.2964    59.5  < 2e-16 ***
## r    0.3022     0.0225    13.4  6.1e-15 ***
## ---
## Signif. codes:
```

```
## 0 '***' 0.001 '**' 0.01 '*' 0.05 '.' 0.1 ' ' 1
##
## Residual standard error: 35.1 on 33 degrees of freedom
##
## Number of iterations to convergence: 6
## Achieved convergence tolerance: 2.44e-06
```

We obtained a converged solution in only 6 iterations (as shown in the output *Number of iterations to convergence: 6*) due to the educated guesstimates of initial values. With this fitted model, the estimated carrying capacity for this mobile banking app is $\hat{C} = 684$ million with a population growth rate of $\hat{r} = 30.2\%$. The time at change point is estimated at the year $\hat{t}_0 = 17.6$. All the estimated parameters are statistically significant with p-values < 0.001. The estimated residual standard error $\hat{\sigma} = 35.07$ with 33 degrees of freedom.

With this satisfactory model fitting, we can predict the customer population and overlay the fitted and predicted population for the observed customer population. In order to see the consequence of using the linear model, we also fit the linear regression and overlaid the linear model fits to this figure. The following *R* code chunk can be used to display this illustration.

```
par(mar = c(4, 4, .1, .1))
# Plot the data
plot(At ~ Month, xlim= c(1,40),
     ylim=c(0, 700), data=dMobile,
     xlab="Month", ylab="Total Customer Size")
# Add the model fitted/predicted
lines(dMobile$Month, predict(Mobile.mod),
      lwd=3)
# Fit the linear regression
lm.fit = lm(At~Month, data=dMobile)
# Print the model fit
summary(lm.fit)
```

```
##
## Call:
## lm(formula = At ~ Month, data = dMobile)
##
## Residuals:
##     Min      1Q  Median      3Q     Max
## -161.35  -62.18    7.92   66.33  169.81
##
## Coefficients:
##             Estimate Std. Error t value Pr(>|t|)
## (Intercept)   -99.70      27.46   -3.63  0.00092 ***
```

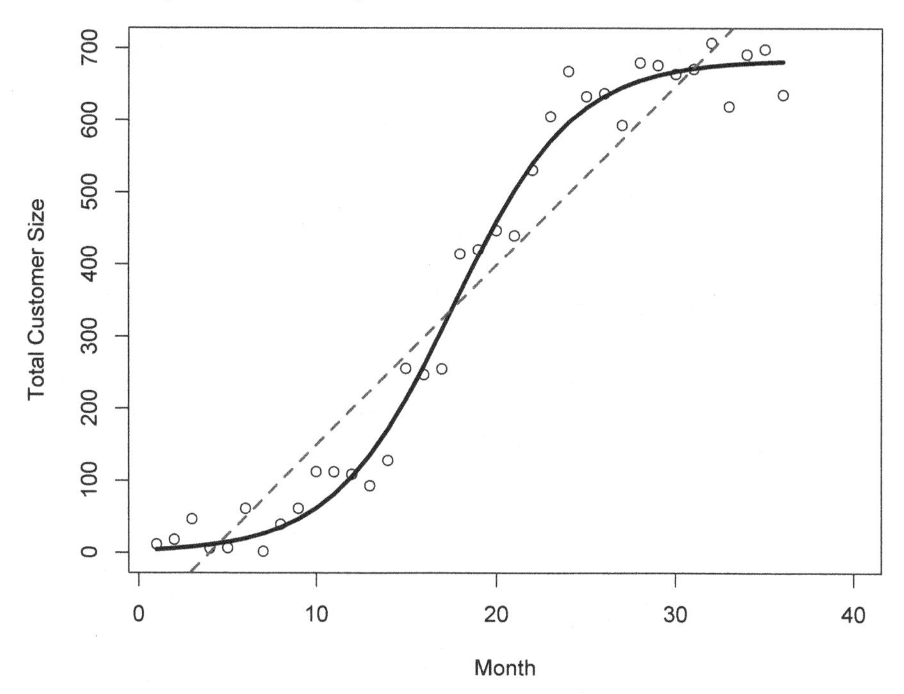

FIGURE 4.3
Illustration of Mobile Banking Population Growth Model

```
## Month          24.91       1.29   19.25  < 2e-16 ***
## ---
## Signif. codes:
## 0 '***' 0.001 '**' 0.01 '*' 0.05 '.' 0.1 ' ' 1
##
## Residual standard error: 80.7 on 34 degrees of freedom
## Multiple R-squared:  0.916,  Adjusted R-squared:  0.913
## F-statistic:  371 on 1 and 34 DF,  p-value: <2e-16
```

```
# Overlay the model fitting
abline(lm.fit, lwd=2, lty=8, col="red")
```

From the above plot in Figure 4.3, it's evident that the observed data was effectively modeled using nonlinear regression through the *R nls* function. This methodology fit the characteristics of the data well, which follow a logistic growth curve marked by distinct growth rates over time. Specifically, the initial phase displays a slow and gradual growth, followed by a rapid expansion, and finally, a tapering off toward the end. The success of the *nls* technique is unsurprising, given its flexibility in handling varying growth rates.

Conversely, the linear regression model fitted by *lm* could not capture the intricacies of the logistic growth curve. Linear regression assumes a consistent growth rate throughout, which isn't suitable for cases where growth rates fluctuate. As such, it's not unexpected that the linear model failed to accurately depict the observed changing growth dynamics from the data.

We can further investigate this misfitting with a Monte-Carlo simulation study.

4.3.3 Possible Consequences Shown in *dMobile* Data

As observed above from the data analysis, using a linear model to fit a nonlinear logistic growth model can lead to several consequences and inaccuracies in the model's performance. These can be summarized in the following points:

1. *Misfitting Data*: Linear regression assumes a constant growth relationship between variables, which is not appropriate for capturing the changing growth rates of a nonlinear logistic growth curve. As a result, the linear model may not accurately capture the observed data points, leading to poor model fit as shown in Figure 4.3.
2. *Biased Parameter Estimates*: Linear regression tries to fit a linear relationship to nonlinear data, which can result in biased estimates for model parameters. This can lead to incorrect interpretations of the growth dynamics and relationships between variables. In the *dMobile* data, the parameters estimated from the nonlinear logistic growth model are meaningful in financial application that the estimated carrying capacity for this mobile banking app is $\hat{C} = 684$ million with a population growth rate of $\hat{r} = 30.2\%$. The time at change point is estimated at the year $\hat{t}_0 = 17.6$. However, the estimated parameters from the linear regression are $\left(\hat{\beta}_0, \hat{\beta}_1\right) = (-99.695, 24.914)$. This can be interpreted as the size of customers in the beginning of this mobile app implementation is -99.695 million and the monthly growth is 24.914 million customers, with this monthly growth continuing over time. This is contradictory to common sense as well as financial and practical knowledge.
3. *Inaccurate Predictions*: If the linear model is used to make predictions beyond the range of the observed data, the predictions can be highly inaccurate. The model won't account for the expected changes in growth rates and will likely make predictions that deviate significantly from the actual values, which can be seen in Figure 4.3.
4. *Poor Extrapolation*: Linear models are not well-suited for extrapolation, especially in cases of nonlinear growth. Extrapolating beyond the range of observed data using a linear model may lead to

unrealistic predictions and incorrect conclusions, which is obvious from Figure 4.3.

5. *Loss of Information*: A linear model's inability to capture the varying growth rates means that important information about the dynamics of financial system is lost. This can hinder the ability to understand underlying processes and make meaningful inferences.

Using the *dMobile* example, we have numerically illustrated the possible consequences on *Misfitting Data*, *Biased Parameter Estimates*, *Inaccurate Predictions*, *Poor Extrapolation*, and *Loss of Information* as shown in Figure 4.3.

For further investigation, we will design and conduct a more extensive Monte-Carlo simulation to further show the consequences of *Inaccurate Predictions* and *Poor Extrapolation*.

4.4 Monte-Carlo Simulation-Based Investigation

4.4.1 Simulation Setup

We make use of the three-parameter logistic growth model (4.4) to simulate data in parallel to the *dMobile* data and evaluate these two models using the predicted residual sum of squares (RSS).

The *simulation steps* are set up as follows:

1. *Truth*: we assume the true parameters (C, r, t_0, σ) = (688, 0.3, 18, 6).
2. *Data Generation*: With the true parameters from *Step 1*, we simulate data from the three-parameter logistic growth model (4.4) with the same time structure from Month 1 to Month 36, i.e., $t = 1, \cdots, 36$. This would generate data $A(i), i = 1, \cdots, 36$ with 36 observations.
3. *Estimation from nls*: With the data generated from *Step 2*, we fit the nonlinear logistic model using *nls* to estimate the model parameters $\left(\hat{C}, \hat{r}, \hat{t}_0, \hat{\sigma}\right)$ and then calculate the predicted values, denoted by $\widehat{A(i)}_{nls}$, based on this fitted nonlinear logistic growth model.
4. *Estimation from lm*: With the data generated from *Step 2*, we fit the linear model using *lm* to estimate the model parameters and then calculate the predicted values, denoted by $\widehat{A(i)}_{lm}$, based on this fitted linear model.

To evaluate the performance and consequences between a nonlinear logistic growth model fitted by *nls* and the linear model fitted by *lm*, we make use of the predicted residual sum of squares (RSS) to compare these two models to the truth.

The *evaluation steps* are set up as follows:

1. *RSS from the True Model*: We calculate the $A(i)$, denoted by $A(i)_T$, using the three-parameter logistic growth model in equation (4.3) with the true parameter values: $(C, r, t_0) = (688, 0.3, 18)$. The RSS can then be calculated by $RSS_T = \sum_i [A(i)_T - A(i)]^2$.
2. *RSS from the Fitted Nonlinear Logistic Growth Model*: We use the predicted $\widehat{A(i)}_{nls}$ from the fitted three-parameter logistic growth model from *nls*. The associated RSS can then be calculated by $RSS_{nls} = \sum_i \left[\widehat{A(i)}_{nls} - A(i)\right]^2$.
3. *RSS from the Fitted Linear Model*: We use the predicted $\widehat{A(i)}_{lm}$ from the fitted linear model from *lm*. The associated RSS can then be calculated by $RSS_{lm} = \sum_i \left[\widehat{A(i)}_{lm} - A(i)\right]^2$.
4. *Ratios of RSS*: We can then calculate the ratios between the RSS for these two models and the RSS from the true model in *Step 1* as $R_{nls} = \frac{RSS_{nls}}{RSS_T}$ for the nonlinear logistic growth model and $R_{lm} = \frac{RSS_{lm}}{RSS_T}$ for the linear model.
5. *Performance Evaluation*: Better performance should then be seen as whichever is close to one. The evaluation of *Inaccurate Predictions* and *Poor Extrapolation* depends on whether i is within (i.e., *prediction*) or outside (i.e., *extrapolation*) the observed time frame of $t = 1, \cdots, 36$.

4.4.2 Numerical Investigation with Monte-Carlo Simulations

We will illustrate a single simulation in step-by-step fashion with detailed explanations. To comprehensively investigate the performance of these two models, we can expand these steps to conduct multiple simulations across a significant number of iterations, such as 10,000. This can be easily implemented using the *R* code developed below with a loop over a large number of times. We leave this exercise to interested readers.

The *simulation steps* on data generation can be implemented in the following *R* code chunk. Note that we will generate 36 months of data (denoted by *data.fit*) for more fitting to evaluate the model fitting performance and 6 more months of data (denoted by *data.ext*) besides the 36 months to evaluate the performance of the two models used for extrapolation. Therefore the entire dataframe (denoted by *data*) has 42 observations. The following *R* code chunk can be used to generate the data:

```
# Set the random seed
set.seed(3388)
# The true parameters in Step 1
C=688; r=0.3; t0=18; sigma=6
# The time frame:
#  Note: 36 months for model fitting
#        and 6 extra months for extrapolation
n=42; tM = 1:n
# The true A(t)
Ai_T = C/(1 + exp(-r * (tM - t0)));
# The observed A(t) in Step 2
eps = rnorm(n, mean=0, sd=sigma)
Ai= Ai_T+eps;
# Make a data frame
data = data.frame(tM= tM, True_Ai = Ai_T, obs_Ai = Ai)
data.fit = data[1:36,] # data for fitting
data.ext = data[37:42,] # data for extrapolation
# Check the data dimension
dim(data);dim(data.fit);dim(data.ext)
```

```
## [1] 42  3

## [1] 36  3

## [1] 6 3
```

```
# Print the first 6 obs
head(data)
```

```
##   tM True_Ai obs_Ai
## 1  1   4.169 14.177
## 2  2   5.616  6.152
## 3  3   7.559 16.084
## 4  4  10.165 14.089
## 5  5  13.650 14.319
## 6  6  18.299 20.514
```

With the generated dataframe *data*, we can now fit the two models as follows:

```
# Fit the nls in Step 3:
nls.fit = nls(obs_Ai~C/(1+exp(-r*(tM-t0))),
              data=data.fit,
```

```
              start=list(C=700, t0=15,r=0.3))
# Print the model fit
summary(nls.fit)

##
## Formula: obs_Ai ~ C/(1 + exp(-r * (tM - t0)))
##
## Parameters:
##    Estimate Std. Error t value Pr(>|t|)
## C  6.92e+02   2.32e+00     298   <2e-16 ***
## t0 1.81e+01   5.13e-02     352   <2e-16 ***
## r  2.95e-01   3.69e-03      80   <2e-16 ***
## ---
## Signif. codes:
## 0 '***' 0.001 '**' 0.01 '*' 0.05 '.' 0.1 ' ' 1
##
## Residual standard error: 5.94 on 33 degrees of freedom
##
## Number of iterations to convergence: 5
## Achieved convergence tolerance: 7.9e-06

# Fit the linear model in Step 4
lm.fit = lm(obs_Ai~tM, data=data.fit)
# Print the model fit
summary(lm.fit)

##
## Call:
## lm(formula = obs_Ai ~ tM, data = data.fit)
##
## Residuals:
##     Min      1Q  Median      3Q     Max
## -113.52  -61.15   -2.92   62.22  105.38
##
## Coefficients:
##             Estimate Std. Error t value Pr(>|t|)
## (Intercept)   -116.7       23.3   -5.02  1.6e-05 ***
## tM              25.5        1.1   23.25  < 2e-16 ***
## ---
## Signif. codes:
## 0 '***' 0.001 '**' 0.01 '*' 0.05 '.' 0.1 ' ' 1
##
## Residual standard error: 68.4 on 34 degrees of freedom
```

```
## Multiple R-squared:  0.941,  Adjusted R-squared:  0.939
## F-statistic:  540 on 1 and 34 DF,  p-value: <2e-16
```

The *evaluation steps* can be implemented in the following *R* code chunk. Note that the prediction inaccuracy and extrapolation are different from the two parts of data where *data.fit* is for fitting and prediction and *data.ext* is for model extrapolation.

We then first evaluate the model fitting and prediction as follows:

```
# Step 1: RSS from the True Model
RSS_T = sum((data.fit$True_Ai-data.fit$obs_Ai)^2)
# Step 2: RSS from the Fitted Nonlinear Logistic Growth Model
pred.nls = predict(nls.fit);
RSS_nls  = sum((pred.nls-data.fit$obs_Ai)^2)
# Step 3: RSS from the Fitted Linear Model
pred.lm = predict(lm.fit)
RSS_lm  = sum((pred.lm-data.fit$obs_Ai)^2)
# Step 4: Ratios of RSS for prediction inaccuracy
R_nls = RSS_nls/RSS_T;R_nls
```

```
## [1] 0.9131
```

```
R_lm  = RSS_lm/RSS_T;R_lm
```

```
## [1] 124.8
```

As seen from the output, the ratio between the *RSS* from the fitted nonlinear logistic growth model to the *RSS* from the truth is 0.9131448, which is close to 1 indicating the fitted nonlinear logistic growth model can be well used to predict the nonlinear relationship. However, the ratio of the *RSS* from the fitted linear model to the *RSS* from the truth is 124.7884, which is far away the required 1, indicating the *prediction inaccuracy* of the fitted linear model for predicting the nonlinear relationship.

We can then evaluate the fitted model for extrapolation to use the *data.ext*, which can be implemented in the following *R* code chunk:

```
# Step 1: RSS from the True Model
RSS_T  = sum((data.ext$True_Ai-data.ext$obs_Ai)^2)
# Step 2: RSS from the Fitted Nonlinear Logistic Growth Model
nls.par  = coef(nls.fit) # get the fitted parameters
C= nls.par[1];t0=nls.par[2];r=nls.par[3]
pred.nls = C/(1 + exp(-r * (data.ext$tM - t0)))
RSS_nls  = sum((pred.nls-data.ext$obs_Ai)^2)
```

```
# Step 3: RSS from the Fitted Linear Model
pred.lm  = predict(lm.fit, new.data=data.ext$tM)
RSS_lm   = sum((pred.lm-data.ext$obs_Ai)^2)
# Step 4: Ratios of RSS for prediction inaccuracy
R_nls = RSS_nls/RSS_T;R_nls
```

```
## [1] 3.388
```

```
R_lm  = RSS_lm/RSS_T;R_lm
```

```
## [1] 85671
```

This is to say that if the fitted nonlinear logistic growth model is used to predict and extrapolate the performance outside the 36 months time frame, the ratio of RSS from the fitted nonlinear logistic growth model to the RSS from the truth is 3.388242, where the ratio of RSS from the fitted linear model would be 85671.23. This shows the *poor extrapolation* if a linear model is fitted and used for extrapolation and prediction whenever the underlying relationship is nonlinear.

4.5 Discussions

Nonlinear models are commonly seen in financial data analysis. There are many nonlinear models to be used in financial modeling.

We illustrate the nonlinear logistic growth models in financial data analysis in this chapter to describe the market penetration of financial products. In this case, the logistic growth model can be applied to describe the adoption and saturation of financial products or services. We considered a new mobile banking app introduced by a bank in this chapter as an example. Initially, the adoption was slow, but as users recognized its benefits and word spread, the adoption rate accelerated. Over time, as most potential users adopted the app, the growth leveled off.

The logistic growth model described in this chapter can be also used for other financial applications, for example

1. *Fintech startups*: For startups in the fintech sector, the logistic growth model can provide insights into the pace of customer acquisition. It helps in understanding when to expect rapid growth and when the growth is likely to stabilize.
2. *Market Share and Competition*: In competitive markets, the logistic growth model can predict the eventual market share each competitor

will achieve. It helps in understanding how the market will be divided among players.

3. *Cryptocurrency Adoption*: Logistic growth can be useful for modeling the adoption of cryptocurrencies. Initially, adoption might be slow, but as more people become familiar with the technology and its benefits, adoption accelerates before eventually reaching a maximum level.

4. *New Financial Services*: When a new financial service, such as robo-advisory platforms, is introduced, the logistic growth model can predict how quickly it will be embraced by investors and when the market will reach a state of maturity.

5. *Insurance Premiums and Policyholders*: The model can help predict the growth of insurance policyholders for a new insurance product, factoring in factors like awareness campaigns and economic conditions.

There are several benefits and considerations in applying logistic growth model:

1. *Strategic Planning*: The logistic growth model aids in strategic planning, allowing companies to allocate resources effectively and set realistic growth expectations.

2. *Market Timing*: It assists in understanding when a market might reach saturation, helping companies plan for the next growth phase or diversification.

3. *Risk Management*: Predicting adoption curves helps manage risks associated with investments in new products. Understanding potential saturation levels is essential for risk assessment.

However, it's important to note that while the logistic growth model provides valuable insights, its application should be done thoughtfully. Real-world adoption can be influenced by various external factors, and the model's assumptions might not always hold true. A combination of domain knowledge, data analysis, and a clear understanding of the specific financial context is crucial for accurate interpretation and decision-making.

4.6 Exercises

1. This exercise is to test the effect of different initial values of the nonlinear model fitting using *nls*. We have used initial values of C

= *700, t0 = 15, r = 0.3* in this chapter to fit the data based on the plot:

```
# Nonlinear regresison with different starting values
Mobile.mod = nls(At~C/(1+exp(-r*(Month-t0))),
                 data=dMobile, start=list(C=700,
                 t0=15,r=0.3))
#Print the model fit
summary(Mobile.mod)
```

Using this *R* code chunk and refit the model using (C, t0, r) = (1000, 50, 0.8) and other sets of initial values.

2. This exercise is to investigate a different setting on the impact of *linear regression* discussed in Section 4.4 using true parameters (C, r, t_0, σ) = (688, 0.3, 18, 30). Notice that the error standard deviation is expanded from $\sigma = 6$ to $\sigma = 30$ to mask the true relationship. Use this set of true values and re-run the simulation in this section to see what you can find and conclude.

5

The Logistic Regression

In the previous chapters, we focused on linear regression with continuous outcomes in financial data analysis which are normally distributed. Not all financial data are normally distributed and we now switch to data analysis for non-normal data in this chapter with focus on binary/binomial outcomes.

Binary outcomes are variables that can take only two values, often denoted as 0 and 1. These are often used to model financial events with a yes/no or success/failure nature. A common example is modeling the success of a stock market in each day trading where 1 representing *gain* and 0 representing *loss*. In comparison, binomial outcomes are more general than the binary outcomes, which are from the binomial probability distribution and are used to model the number of successes in a fixed number of independent Bernoulli trials. A common example is to modeling the number of successful days in a month(or in a year) (i.e., fixed number of Bernoulli trials in a fixed number of time in a year).

The logistic regression is a common method for modeling binary/binomial outcomes, where the logistic function transforms a linear combination of predictors into probabilities. The logistic regression models are part of the *generalized linear model* (*GLM*) which has been widely introduced and well explained, see for example McCullagh and Nelder (1995), Dobson and Barnett (2018), and Dunn and Smyth (2018). Interested readers can refer to these publications for detailed theoretical development and in this chapter, we will briefly describe the logistic regression and concentrate on using *R* function *glm* to analyze a dataset called *Smarket* from *R* package *ISLR2*, which is created for the book by James et al. (2023) with title *Introduction to Statistical Learning with applications in R, Second Edition.*

5.1 The Logistic Regression

5.1.1 Common Misconceptions

Let's start by clarifying some common misconceptions in the analysis of binary/binomial data. This will guide us toward the correct modeling techniques

DOI: 10.1201/9781003469704-5

in maximum likelihood estimation for these types of data. A more comprehensive Monte-Carlo simulation-based validation study will be conducted in Section 5.3.

5.1.1.1 Misconception about Linear Regression

The linear regression techniques (*least squares*) discussed in the preceding chapters are no longer applicable for modeling dichotomous outcome variables.

For examples with binary outcomes, the outcome variable y can only take values of 0 (i.e., stock market at loss) or 1 (i.e., stock market at gain) with one or more independent variables (continuous or categorical) of $(x_1, \cdots, x_K)$, the classical multiple linear regression to the K independent variables:

$$y = \beta_0 + \beta_1 x_1 + \cdots + \beta_K x_K + \epsilon \tag{5.1}$$

is not appropriate anymore due to the binary outcome values of 0 and 1. With binary outcomes, the error term ϵ can not be assumed to be normally distributed anymore. In addition, the assumption about constant variance (i.e., homoscedasticity) can not be held anymore since the mean and variance are linked to each other for binary data. These violations on the classical assumptions for multiple linear regression will automatically invalidate the conclusions if we use it to analyze binary data.

5.1.1.2 Misconception about Logit-Transformed Linear Regression

Another common misconception in modeling binary/binomial data is to model the probability with a logit transformation as follows:

$$log\left(\frac{p}{1-p}\right) = \beta_0 + \beta_1 x_1 + \cdots + \beta_K x_K + \epsilon \tag{5.2}$$

where $p = Prob(Y = 1)$. This misconception is generated due to the fact that the correct logistic regression can be written in the following form after estimation via the theory of maximum likelihood:

$$log\left(\frac{\hat{p}}{1-\hat{p}}\right) = \hat{\beta}_0 + \hat{\beta}_1 x_1 + \cdots + \hat{\beta}_K x_K. \tag{5.3}$$

Both equations in (5.2) and (5.3) look similar, but logically they are very different. The equation (5.3) is the estimated model after maximum likelihood estimation, which we will illustrate later, but the equation (5.2) is methodologically and logically incorrect due to two reasons. The first reason is that the error term ϵ in equation (5.2) can not be assumed to follow the multiple linear regression as described in the misconception subsection (5.1.1.1). The second and most important reason is that the left-side $log\left(\frac{p}{1-p}\right)$ in equation (5.2) is not defined since we don't know $p = Prob(Y = 1)$ so least squares estimation can not be performed.

5.1.2 Logistic Regression with MLE

5.1.2.1 Binomial Data

As discussed above, the least squares estimation using for multiple linear regression is not appropriate anymore to model binary/binomial data. In this section, we briefly describe the foundation of logistic regression built on the maximum likelihood estimation with binomial data since binary data is a special case of binomial data.

Let's denote $Y_i \sim B(n_i, p_i)$ as the financial binomial response variable Y_i ($i = 1, \cdots, n_i$, and n_i fixed) with the associated probability p_i. Note that this binomial variable Y_i becomes the binary variable if $n_i = 1$. To assist with understanding, think about Y_i as the number of days of successful stock trading in a month (let's assume that there are 30 days in a month at this moment, i.e., $n_i = 30$) so Y_i can take any values between 0 and 30. If $n_i = 1$ to restrict the trading on each day, then Y_i would be just 0 (loss) or 1 (success). The probability distribution can then be written as:

$$P(Y_i = y_i) = \binom{n_i}{y_i} p_i^{y_i}(1 - p_i)^{n_i - y_i} \tag{5.4}$$

The binomial response variable Y_i and the associated probability p_i may be related to K covariates, such as, investment firm's stock shares, employee's experience, etc., which are denoted by $(x_{i1}, \cdots, x_{iK})$. The fundamental difference between logistic regression and multiple linear regression is that the response variable (i.e., outcome variable) is binomially distributed, not normally distributed anymore. We then model the probability p_i as a linear function of the K covariates. This is the basic idea behind logistic regression and is described as part of the *generalized linear model (GLM)* developed for the exponential family of distributions.

5.1.2.2 Maximum Likelihood Estimation (MLE)

In the generalized linear model framework, we use *a linear predictor* to model the linear relationship and a *link function* to link the *linear predictor* to the binomial probability p_i. Specifically, the *linear predictor* is denoted by:

$$\eta_i = \beta_0 + \beta_1 x_{i1} + \cdots + \beta_K x_{iK} = X_i\beta \tag{5.5}$$

where $X_i = (1, x_{i1}, \cdots, x_{iK})$ is the matrix of observed covariates and $\beta = (\beta_0, \beta_1, \cdots, \beta_K)$ is the associated parameter vector.

Various *link functions* are possible for linking the *linear predictor* η, which is a linear combination of the effects of one or more explanatory variables, to the (outcome) probabilities p_i that we want to model. It is easy to see that the identity link of $p_i = \eta_i$ is not appropriate since the binomial probability p_i has to be constrained to the [0, 1] interval. For binomial response, the most commonly used link function is the so-called *logit* (and therefore giving rise to the term logistic regression) given by:

$$\eta = log\left(\frac{p}{1-p}\right) \tag{5.6}$$

or

$$p = \frac{e^{\eta}}{1+e^{\eta}} \tag{5.7}$$

where $log\left(\frac{p}{1-p}\right)$ is the so-called log-odds. With this *logit* link, we have a model which can be *symbolically* expressed as $log\left(\frac{p}{1-p}\right) = \beta_0 + \beta_1 x_1 + \cdots + \beta_K x_K$. With this model, the β_0 can be interpreted as the log-odds without the effects of all independent variables $(x_1, x_2, \cdots, x_K)$ and the $\beta_i (i = 1, 2, \cdots, K)$ are the coefficients that indicate the effect of each independent variable x_i on the log odds of the outcome and are interpreted as the log odds-ratio change for every unit change of independent variable x_i.

Other link functions used to model binomial response data are the *probit link function*: $\eta = \Phi^{-1}(p)$, where Φ^{-1} is the inverse normal cumulative distribution function, which produces the *probit regression*.

Maximum likelihood estimation (MLE) is used for parameter estimation and statistical inference. To perform MLE, we first specify the likelihood function to be maximized. The general likelihood function is defined as:

$$P(Y = y) = \prod_{i=1}^{n} f(y_i|\theta) \equiv (\theta|y) \tag{5.8}$$

For binomial data, the likelihood function becomes:

$$L(\beta|y) = \prod_{i=1}^{n} \binom{n_i}{y_i} p_i^{y_i} (1-p_i)^{n_i - y_i} \tag{5.9}$$

The likelihood is a function of the unknown parameters, the observed response data, and the covariates. MLE of the unknown parameters requires finding values of the parameters that maximize the likelihood function. Maximizing the likelihood function is equivalent to maximizing the log-likelihood function:

$$l(\theta|y) = logL(\theta|y). \tag{5.10}$$

For binomial data:

$$l(\beta|y) = \sum_{i=1}^{n} \left[log\binom{n_i}{y_i} + y_i log(p_i) + (n_i - y_i) log(1-p_i) \right]. \tag{5.11}$$

If we use the *logit link function* of $\eta = log\left(\frac{p}{1-p}\right)$, then $p = \frac{e^\eta}{1+e^\eta}$. The log-likelihood function becomes:

$$\begin{aligned} l(\beta|y) &= \sum_{i=1}^{n}\left[log\binom{n_i}{y_i} + y_i log(p_i) + (n_i - y_i)log(1-p_i)\right] \\ &= \sum_{i=1}^{n}\left[y_i\eta_i - n_i log(1+e^{\eta_i}) + log\binom{n_i}{y_i}\right] \end{aligned} \tag{5.12}$$

This function is then maximized to obtain the parameter estimates. In software implementation for optimization, minimization is typically implemented and therefore the theory of maximum likelihood estimation to maximize the log-likelihood function is in fact to minimize the *negative* log-likelihood function.

There is no analytical closed-form solution for the parameter estimates as there was in normal-distribution-based multiple linear regression in Chapter 2. Numerical search methods are required and maximum likelihood estimation theory is drawn upon for parameter estimates and their standard errors, confidence intervals, p-values as well as model selection. This logistic regression is implemented in *R* as a function *glm*, which will be illustrated in the following section.

5.2 S&P Stock Market Data Analysis

5.2.1 Descriptive Data Analysis

To illustrate logistic regression using *R* function *glm*, we will use a dataset called *Smarket* from *R* package *ISLR2*, which is created for the book by James et al. (2023) with title *Introduction to Statistical Learning with applications in R, Second Edition.*

The dataset *Smarket* is about the daily percentage returns for the S&P 500 stock index between 2001 and 2005, which is compiled from the raw values of the S&P 500 obtained from Yahoo Finance and then converted to percentages and lagged. It has 1,250 observations on the following 9 variables:

- *Year*: The year that the observation was recorded,
- *Lag1*: Percentage return for previous day,
- *Lag2*: Percentage return for 2 days previous,
- *Lag3*: Percentage return for 3 days previous,
- *Lag4*: Percentage return for 4 days previous

- *Lag5*: Percentage return for 5 days previous
- *Volume*: Volume of shares traded (number of daily shares traded in billions)
- *Today*: Percentage return for today
- *Direction*: A factor with levels Down and Up indicating whether the market had a positive or negative return on a given day

Let's look at the data:

```
# load the library which includes the data
library(ISLR2)
# load the data into R and check the dimension
data(Smarket); dim(Smarket)
```

```
## [1] 1250    9
```

```
# print the summary of the data
summary(Smarket)
```

```
##       Year           Lag1               Lag2
##  Min.   :2001   Min.   :-4.922   Min.   :-4.922
##  1st Qu.:2002   1st Qu.:-0.639   1st Qu.:-0.639
##  Median :2003   Median : 0.039   Median : 0.039
##  Mean   :2003   Mean   : 0.004   Mean   : 0.004
##  3rd Qu.:2004   3rd Qu.: 0.597   3rd Qu.: 0.597
##  Max.   :2005   Max.   : 5.733   Max.   : 5.733
##       Lag3               Lag4               Lag5
##  Min.   :-4.922   Min.   :-4.922   Min.   :-4.922
##  1st Qu.:-0.640   1st Qu.:-0.640   1st Qu.:-0.640
##  Median : 0.038   Median : 0.038   Median : 0.038
##  Mean   : 0.002   Mean   : 0.002   Mean   : 0.006
##  3rd Qu.: 0.597   3rd Qu.: 0.597   3rd Qu.: 0.597
##  Max.   : 5.733   Max.   : 5.733   Max.   : 5.733
##      Volume           Today           Direction
##  Min.   :0.356   Min.   :-4.922   Down:602
##  1st Qu.:1.257   1st Qu.:-0.639   Up  :648
##  Median :1.423   Median : 0.038
##  Mean   :1.478   Mean   : 0.003
##  3rd Qu.:1.642   3rd Qu.: 0.597
##  Max.   :3.152   Max.   : 5.733
```

```
# print the first 8 observations for example
head(Smarket, n=8)
```

```
# Year  Lag1   Lag2   Lag3   Lag4   Lag5 Volume Today Direction
# 2001  0.381 -0.192 -2.624 -1.055  5.010 1.191  0.959        Up
# 2001  0.959  0.381 -0.192 -2.624 -1.055 1.296  1.032        Up
# 2001  1.032  0.959  0.381 -0.192 -2.624 1.411 -0.623      Down
# 2001 -0.623  1.032  0.959  0.381 -0.192 1.276  0.614        Up
# 2001  0.614 -0.623  1.032  0.959  0.381 1.206  0.213        Up
# 2001  0.213  0.614 -0.623  1.032  0.959 1.349  1.392        Up
# 2001  1.392  0.213  0.614 -0.623  1.032 1.445 -0.403      Down
# 2001 -0.403  1.392  0.213  0.614 -0.623 1.408  0.027        Up
```

As seen from the output, this data set consists of percentage returns for the *S&P 500 stock index* over 1,250 days, from the start of 2001 to the end of 2005. For each day, there are data on the percentage returns for each of the five previous trading days denoted by *Lag1* through *Lag5*, the recorded *Volume* as the number of shares traded on the previous day (in billions), *Today* as the percentage return on the date in question and *Direction* as on whether the market was *Up* or *Down* on that day. Note that the *Direction* is a qualitative variable closely related to the percentage return of *Today*. If *Today* is positive, the *Direction* is *Up*, otherwise it is *Down*.

From the financial market theory, we expect to see that the returns from the previous days should be correlated with the return of *Today* and we can easily verify that using the *R* function *cor()* to calculate all of the pairwise correlations among the predictors in the data set using the following *R* code chunk:

```
# Exclude the qualitative variable: Direction
#   and round the correlation to the 3rd digit
round(cor(data.frame(Smarket[,1:8])),3)
```

```
##          Year   Lag1   Lag2   Lag3   Lag4   Lag5 Volume  Today
## Year    1.000  0.030  0.031  0.033  0.036  0.030  0.539  0.030
## Lag1    0.030  1.000 -0.026 -0.011 -0.003 -0.006  0.041 -0.026
## Lag2    0.031 -0.026  1.000 -0.026 -0.011 -0.004 -0.043 -0.010
## Lag3    0.033 -0.011 -0.026  1.000 -0.024 -0.019 -0.042 -0.002
## Lag4    0.036 -0.003 -0.011 -0.024  1.000 -0.027 -0.048 -0.007
## Lag5    0.030 -0.006 -0.004 -0.019 -0.027  1.000 -0.022 -0.035
## Volume 0.539  0.041 -0.043 -0.042 -0.048 -0.022  1.000  0.015
## Today  0.030 -0.026 -0.010 -0.002 -0.007 -0.035  0.015  1.000
```

As seen from the correlations, all the correlations between the lag variables (*Lag1* to *Lag5*) and today's returns (*Today*) are close to zero (the last row or last column). This is also true for all the correlations among all the lagged variables themselves. In other words, there appears to be little correlation between today's returns and previous days' returns as well as among all the

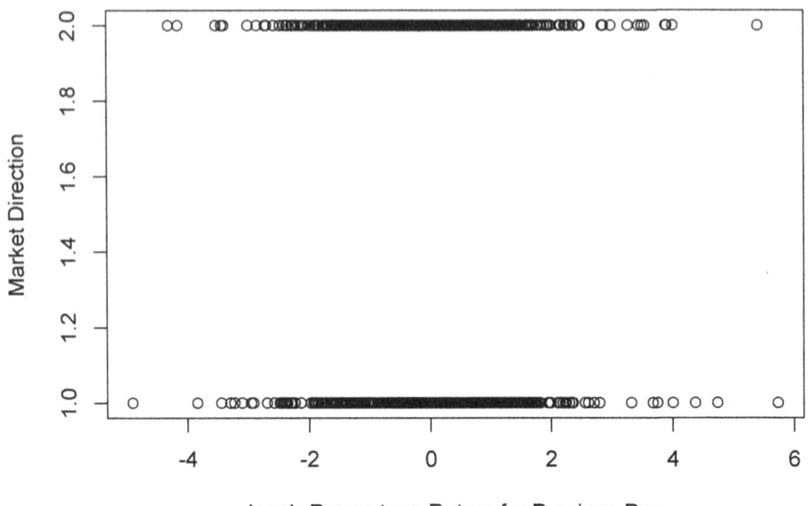

FIGURE 5.1
Relationship between Lag1 and Direction

lagged variables. The only substantial correlation between *Year* and *Volume* is 0.539 which helps little in predicting the stock's daily movements.

This observation can be graphically illustrated using the following plot between stock *Direction* and the lagged variables (only *Lag1* illustrated here and interested readers can plot it with other lagged variables):

```
plot(as.numeric(Direction)~Lag1, Smarket,
     xlab="Lag1: Percentage Return for Previous Day",
     ylab="Market Direction")
```

As seen in Figure 5.1, the stock *Direction* is clustered around zero for both *UP/Down* and there is no distinction as increasing in *Lag1*: Percentage return for previous day, indicating that stock *Direction* is *not really* related to the *Lag1*.

5.2.2 Model Fitting with Logistic Regression

For a more comprehensive analysis, we now come to use logistic regression to model the stock *Direction* either *UP/Down* with the other stock variables of *Lag1* to *Lag5* and *Volume*. Logistic regression models can be fitted easily in *R* using the *glm*. Since *glm* can be used for different distributions, we need to specify the associated probability distribution, i.e., *family* = *binomial*, as an argument to tell *R* to perform *logistic regression* with binomial distributed data instead of other *generalized linear models* as we will illustrate in Chapter 6 for counts data.

The syntax of logistic regression is as follows:

```
# Fit logistic regression using glm
modSP.logistic=glm(Direction~Lag1+Lag2+Lag3+Lag4+Lag5+Volume,
    data = Smarket, family = binomial)
# Print the fit
summary(modSP.logistic)
```

```
##
## Call:
## glm(formula = Direction ~ Lag1 + Lag2 + Lag3 + Lag4 + Lag5 +
##     Volume, family = binomial, data = Smarket)
##
## Coefficients:
##             Estimate Std. Error z value Pr(>|z|)
## (Intercept) -0.12600    0.24074   -0.52     0.60
## Lag1        -0.07307    0.05017   -1.46     0.15
## Lag2        -0.04230    0.05009   -0.84     0.40
## Lag3         0.01109    0.04994    0.22     0.82
## Lag4         0.00936    0.04997    0.19     0.85
## Lag5         0.01031    0.04951    0.21     0.83
## Volume       0.13544    0.15836    0.86     0.39
##
## (Dispersion parameter for binomial family taken to be 1)
##
##     Null deviance: 1731.2  on 1249  degrees of freedom
## Residual deviance: 1727.6  on 1243  degrees of freedom
## AIC: 1742
##
## Number of Fisher Scoring iterations: 3
```

As seen from the output of this logistic modeling, all 6 independent valuables are not statistically significant as shown by their p-values in the column *Pr(>|z|)*. The smallest p-value is associated with *Lag1* with a negative coefficient indicating that if the market had a positive return in the previous day, it is less likely to go up today. However, this is not statistically significant since the associated p-value is at 0.145, which is quite large indicating there isn't statistically significant evidence for a real association between *Lag1* and stock *Direction* today.

This fact can be further strengthened using the *chi-square likelihood ratio test*. In logistic regression, we examine the *deviance* value to assess the model fit to the data. *Deviance* is a measure of goodness of fit of a generalized linear model. Or rather, it is a measure of *badness* of fit, with higher values indicating a worse or inaccurate fit to the data. As seen from the output, *R* reports two types of deviance - the *null deviance* and the *residual deviance.* The *null deviance* shows how well the response variable (i.e., *Direction* in this example) is predicted by a model that includes only the intercept (grand mean). In this example, we have a value of 1731.2 on 1249 degrees of freedom. Including the other 6 independent variables decreased the deviance (now it is denoted by the *residual deviance*) to 1727.6 on 1243 degrees of freedom, we can employ the *likelihood ratio test(LRT)* to see whether this reduction is statistically significant or not using the following *R* code chunk:

```
# Extract the null deviance
dev.null   = modSP.logistic$null.deviance;
dev.null

## [1] 1731

# Extract the residual deviance
dev.resid  = deviance(modSP.logistic);
dev.resid

## [1] 1728

# Extract the dfs
df.null    = modSP.logistic$df.null;
df.null

## [1] 1249

df.resid  = modSP.logistic$df.residual;
df.resid

## [1] 1243

# Calculate the p-value from the
#   chi-square likelihood ratio test
pval = 1-pchisq(dev.null-dev.resid,
                df.null-df.resid)
# Print the p-values for the LRT
pval

## [1] 0.7319
```

As evident from the computation, the p-value of 0.7319 suggests that the logistic regression with six independent variables lacks statistical significance in predicting the *Direction* of the stock. This outcome aligns with financial theory, which acknowledges the intricate nature of the stock market.

5.2.3 Prediction using Logistic Regression

With the fitted logistic regression model in the above section, we can do some predictions using the *R* function *predict()*. To predict the likelihood of the market direction given certain values of predictors, the fitted logistic regression model can be used to output probabilities in the form of $P(Y = 1/X)$ using the *R* option *type = "response"*. If there is no new data set provided, the *predict()* function will just provide the probabilities based on the data used for the logistic regression modeling. The *R* code chunk is as follows:

```
# Predicted probabilities
modSP.predP = predict(modSP.logistic, type = "response")
# Print the first 10
modSP.predP[1:10]
```

```
##      1      2      3      4      5      6      7
## 0.5071 0.4815 0.4811 0.5152 0.5108 0.5070 0.4927
##      8      9     10
## 0.5092 0.5176 0.4888
```

These predicted probabilities associated with the 6 independent variables are numerical values from 0 to 1, which should be translated to predict if the market will go up or down on a specific day. For that purpose, we need to transform the anticipated probabilities into qualitative labels of *Up* or *Down* to be compared with the observed stock *Direction*. Using the cutoff probability of 0.5 in the middle of interval from 0 to 1 (of course, you can select different cutoff values based on your financial knowledge!), we can create a class variable to describe the predicted stock market direction. The *R* code chunk is as follows:

```
# Create a new class variable: modSP.predD
modSP.predD = rep("UP", nrow(Smarket))
# Make "Down" if p < 0.5
modSP.predD[modSP.predP < 0.5] = "Down"
# Print the first 6
modSP.predD[1:6]
```

```
## [1] "UP"   "Down" "Down" "UP"   "UP"   "UP"
```

With this newly created prediction variable *modSP.predD*, we can compare it with the observed *Direction* using *table()* function to create a *confusion matrix* that shows the number of correct and incorrect classifications as follows:

```
table(modSP.predD, Smarket$Direction)
```

```
##
## modSP.predD Down  Up
##        Down  145 141
##        UP    457 507
```

In the above confusion matrix, the diagonal elements indicate correct predictions and the off-diagonals represent incorrect predictions. Using this fitted logistic regression model, we correctly predicted that the market would go up on 507 days and go down on 145 days, for a total of 507 + 145 = 652 correct predictions, which means that we have 52.16% (652/1250) accuracy in using this logistic regression model to predict the behavior of the stock market. This is no impressive at all and just a little better than *random guessing*, but we know that the stock market is complicated and we can not simply use the previous performance to accurately predict future performance.

5.3 Monte-Carlo Simulation-Based Validation Study

In order to provide a deeper understanding of the common *misconceptions* discussed in Section 5.1.1, we conduct an extensive validation study in this section using Monte-Carlo simulation techniques. The purpose is to offer a more comprehensive exploration of the topic.

To achieve this, we will design a Monte-Carlo simulation that closely mirrors the binary data frequently encountered in financial data analysis. Through this simulation, we will proceed to apply both the *logistic regression model* and the *linear model*. By doing so, we can quantitatively showcase the biases that can arise from employing *linear regression* in situations where *logistic regression* is more appropriate.

This simulation-based approach serves to clarify the *Misconception about Linear Regression* as well as the *Misconception about Logit-Transformed Linear Regression*. By working through these numerical results, we hope to offer interested readers a clearer grasp of the limitations and implications associated with these misconceptions. This step-by-step demonstration with simulated data should serve as a powerful tool for enhancing readers' comprehension of the complex relationship between different regression techniques and their proper applications.

5.3.1 Simulation Design

The design using *R* involves generating artificial binary data based on known parameters and then fitting the logistic regression model and the corresponding linear model to the generated data. Here's a step-by-step guide on how to set up this simulation study:

- *Step 1 to Set Up Parameters*: Define the coefficients and intercept for the logistic regression model. These coefficients determine the relationship between the independent variables and the log-odds of the binary outcome. In this simulation, we show an investment decision for the S&P500 with relationship to the investors' *age* and *income* status. We set *intercept* $= \beta_0 = 0$, which means that the log-odds without the effects of *age* and *income* is 0. With log-odds of 0, the probability to invest or not can be calculated as $p = 50\%$, indicating a 50-50 chance for investment. We also assume that the *coef_age* $= \beta_1 = 0.1$ and *coef_income* $\beta_2 = 0.2$, which corresponds to the odds-ratio of $e^{\beta_1} = e^{0.1} = 1.105$ for *age* and $e^{\beta_2} = e^{0.2} = 1.221$ for *income*.

- *Step 2 to Generate Independent Variables*: Generate data for the two independent variables with $x_1 =$ *age* simulated from a *uniform* distribution from age of 20 to 50 and $x_2 =$ *income* from a normal distribution with mean annual income of $60,000 and standard deviation of $15,000. We *scale* these two independent variables to mean of 0 and standard deviation of 1 using the *R* function *scale* because otherwise the calculated linear predictor $\eta = \beta_0 + \beta_1 x_1 + \beta_2 x_2$ would be too big with the parameters from *Step 1* and the generated *age* and *income*, which would result the simulated p to be all 1.

- *Step 3 to Calculate Log-Odds*: Calculate the log-odds of the binary outcome using the formula: *log-odds* $= \log\left(\frac{p}{1-p}\right) = \eta = \beta_0 + \beta_1 x_1 + \beta_2 x_2$.

- *Step 4 to Transform to Probabilities*: Convert the calculated log-odds to probabilities using the logistic function: $p = \frac{1}{1+exp(-\eta)}$.

- *Step 5 to Generate Binary Outcomes*: For each observation, generate binary outcomes (0s and 1s) based on the calculated probabilities p from *Step 4* using *R* function *rbinom()* to generate binomially distributed random numbers.

- *Step 6 to Fit Logistic Regression Model and Linear Model*: Use the generated data, we can fit the logistic regression model using the *glm()* function and linear model using *lm* function in *R*.

For a comprehensive Monte-Carlo validation study, we can repeat the *Step 1* to *Step 6* for a large number of times and then summarize the parameter estimation from these simulated outputs.

5.3.2 Simulation Implementation in *R*

The above-outlined procedures can be effectively translated into *R* programming language using the following code. We will provide a comprehensive explanation of this code to ensure that readers can understand the steps easily.

```
# Set the random seed for reproducibility
set.seed(123)
# Step 1: Set up parameters
intercept = 0; coef_age = 0.1; coef_income = 0.2
# The number of observation
n = 1000
# Step 2: Generate independent variables and scale
age    = scale(runif(n, 20, 50))
income = scale(rnorm(n, 60000, 15000))
# Step 3: Calculate log-odds
log_odds = intercept+coef_age*age+coef_income*income
# Step 4: Transform to probabilities
prob = 1 / (1 + exp(-log_odds))
# Step 5:  Generate binary outcomes
outcomes = rbinom(n, 1, prob)
# Create a data frame
simu_data = data.frame(age, income, outcomes)
# Print the first 6 observation from the 1000
head(simu_data)
```

```
##       age  income outcomes
## 1 -0.7294 -0.6129        0
## 2  1.0123 -1.0041        1
## 3 -0.3072  1.0133        1
## 4  1.3418  0.7380        0
## 5  1.5416 -1.5188        0
## 6 -1.5713 -0.1069        1
```

```
# Print the dimension of the data
dim(simu_data)
```

```
## [1] 1000    3
```

```
# Step 6.1: Fit logistic regression model
fit.logisticM = glm(outcomes ~ age + income,
                data = simu_data, family = "binomial")
# Display model summary
summary(fit.logisticM)
```

```
##
## Call:
## glm(formula = outcomes ~ age + income, family = "binomial",
## data = simu_data)
##
## Coefficients:
##              Estimate Std. Error z value Pr(>|z|)
## (Intercept)  -0.0162     0.0635   -0.25   0.7993
## age           0.0871     0.0636    1.37   0.1709
## income        0.1677     0.0640    2.62   0.0088 **
## ---
## Signif. codes:
## 0 '***' 0.001 '**' 0.01 '*' 0.05 '.' 0.1 ' ' 1
##
## (Dispersion parameter for binomial family taken to be 1)
##
##     Null deviance: 1386.2  on 999  degrees of freedom
## Residual deviance: 1377.5  on 997  degrees of freedom
## AIC: 1384
##
## Number of Fisher Scoring iterations: 3
```

```
# Step 6.2: Fit linear regression model
fit.lm <- lm(outcomes ~ age + income, data = simu_data)
# Display model summary
summary(fit.lm)
```

```
##
## Call:
## lm(formula = outcomes ~ age + income, data = simu_data)
##
## Residuals:
##    Min     1Q Median     3Q    Max
## -0.618 -0.493 -0.382  0.498  0.649
##
## Coefficients:
##             Estimate Std. Error t value Pr(>|t|)
## (Intercept)   0.4960     0.0158   31.46   <2e-16 ***
## age           0.0216     0.0158    1.37   0.1713
## income        0.0416     0.0158    2.63   0.0086 **
## ---
## Signif. codes:
## 0 '***' 0.001 '**' 0.01 '*' 0.05 '.' 0.1 ' ' 1
##
## Residual standard error: 0.499 on 997 degrees of freedom
```

```
## Multiple R-squared:  0.00865,    Adjusted R-squared:  0.00666
## F-statistic: 4.35 on 2 and 997 DF,  p-value: 0.0132
```

As evidenced by the above *R* code and the resulting output, we have constructed a dataset named *simu_data* comprising 1,000 observations. Inspecting the initial six observations, we can see that: 1) the two independent variables of *age* and *income* have been scaled with a *mean* of 0 and a *standard deviation* of 1, and 2) the dependent variable *outcomes* is a binary variable indicating investment decisions for S&P as either 1 (Yes) or 0 (No).

Upon employing the *glm* function to fit the *logistic regression model*, we obtain parameter estimates $(\hat{\beta}_0, \hat{\beta}_1, \hat{\beta}_2)$ of (-0.01615, 0.08710, 0.16770) and the corresponding standard errors (0.06352, 0.06361, 0.06400). These estimates subsequently yield z-statistics (-0.254, 1.369, 2.620) and associated p-values (0.79930, 0.17092, 0.00879). Notably, in this *logistic regression model*, the only predictor variable that attains significance concerning investment decisions in the S&P context is *income*.

Likewise, conducting a linear regression model via the *lm* function yields estimated parameters $(\hat{\beta}_0, \hat{\beta}_1, \hat{\beta}_2)$ of (0.49600, 0.02160, 0.04156) along with the corresponding standard errors (0.01577, 0.01578, 0.01578). These estimates lead to t-statistics (31.460, 1.369, 2.634) and corresponding p-values (<2e-16, 0.17132, 0.00857). In this linear regression model, similar to the *logistic regression* scenario, *income* emerges as the only predictor with notable influence over investment decisions in the S&P context.

Upon juxtaposing the estimated parameters derived from both the *logistic regression model* and the *linear regression model* against the true parameter values $(\beta_0, \beta_1, \beta_2) = (0, 0.1, 0.2)$, it becomes distinctly apparent that the *logistic regression model* showcases considerably fewer biases when contrasted with its *linear regression* counterpart.

In addition, fitting linear regression model to binary data violates the assumptions of *linear regression* where the residuals should be normally distributed with constant variance. A graphical illustration can be easily constructed for the *residual plot* to plot the *residuals* to the *fitted values* and the associated residual distribution plot using the *QQ-plot*. This can be done using the following *R* code chunk.

```
# Extract the residuals
resid.lm = fit.lm$residuals
# Extract the fitted values
fitted.lm = fit.lm$fitted.values
# Contract the Residual plot
par(mfrow=c(1,2))
plot(resid.lm~fitted.lm, xlab="Fitted Values",
     ylab="Residuals", main="Residual Plot")
```

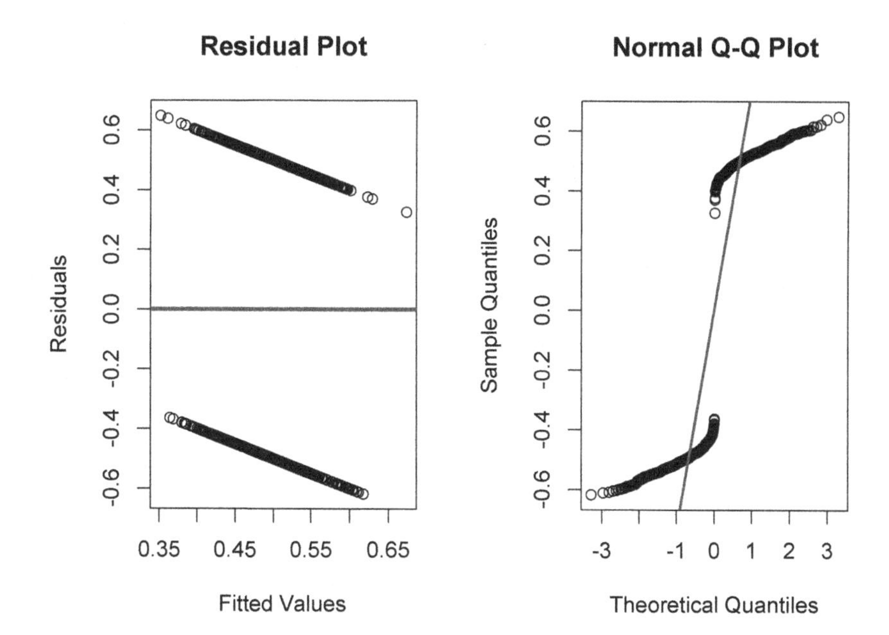

FIGURE 5.2
Residual Diagnostics for Linear Regression Model

```
abline(h=0, lwd=3, col="red")
# Contract the QQ-plot
qqnorm(resid.lm)
qqline(resid.lm, lwd=2, col="red")
```

As shown in Figure 5.2, the *residual plot* in the left panel displays a clear pattern of decreasing residuals which is supposed to be randomly distributed around the mean 0 (i,e., the horizontal *red* line). In addition, the *Normal Q-Q Plot* should have shown that the residual sample quantiles are around the *red* line to be close to the theoretical normal quantiles. This is not the case from both figures in Figure 5.2, indicating that the residuals from the *linear regression model* are not normally distributed with constant variance.

As depicted in Figure 5.2, the *residual plot* on the left panel reveals a discernible pattern characterized by a systematic decrease in residuals. Ideally, these residuals should exhibit random distribution around the mean of 0 as displayed by the horizontal *red* line. Furthermore, the anticipated behavior for the *Normal Q-Q Plot* would involve the residual sample quantiles aligning closely with the *red* line, indicative of a proximity to theoretical normal quantiles. However, this anticipated pattern is conspicuously absent in both visualizations within Figure 5.2. This incongruence serves as an indicator that the residuals resulted from the *linear regression model* deviate from the assumptions of normal distribution with a constant variance.

5.3.3 A Full-Scale Monte-Carlo Simulation Validation.

To systematically investigate the bias from the *liner model*, we can conduct a full-scale Monte-Carlo simulation study to run the 6-step procedure in Section 5.3.1 for a large number of times (i.e., *nsimu=1,000*). This can be implemented with a looping structure in the following *R* code chunk:

```
# Set the random seed for reproducibility
set.seed(123)
# The number of simulations
nsimu=1000
# Create a matrix to hold the simulation results
#   which has 6 columns for the 6 parameters
estParMat = matrix(0, ncol=6, nrow=nsimu)
colnames(estParMat) = c("b0.glm","b1.glm","b2.glm",
                        "b0.lm","b1.lm","b2.lm")
# Now looping
for(s in 1:nsimu){
# Step 1: Set up parameters
intercept = 0; coef_age = 0.1; coef_income = 0.2
# The number of observation
n = 1000
# Step 2: Generate independent variables and scale
age    = scale(runif(n, 20, 50))
income = scale(rnorm(n, 60000, 15000))
# Step 3: Calculate log-odds
log_odds = intercept+coef_age*age+coef_income*income
# Step 4: Transform to probabilities
prob = 1 / (1 + exp(-log_odds))
# Step 5:  Generate binary outcomes
outcomes = rbinom(n, 1, prob)
# Create a data frame
simu_data = data.frame(age, income, outcomes)
# Step 6.1: Fit logistic regression model
fit.logisticM = glm(outcomes ~ age + income,
                    data = simu_data, family = "binomial")
# Track the 3 estimated parms from glm
estParMat[s,1:3] = fit.logisticM$coefficients
# Step 6.2: Fit linear regression model
fit.lm <- lm(outcomes ~ age + income, data = simu_data)
```

```
# Track the 3 estimated parms from lm
estParMat[s,4:6] = fit.lm$coefficients
} # end of s-loop
# Print the summary of the simulated result
dim(estParMat)
```

```
## [1] 1000    6
```

```
summary(estParMat)
```

```
##      b0.glm              b1.glm             b2.glm
##  Min.   :-0.17823   Min.   :-0.1287   Min.   :0.0198
##  1st Qu.:-0.05224   1st Qu.: 0.0544   1st Qu.:0.1597
##  Median :-0.00416   Median : 0.0976   Median :0.2020
##  Mean   :-0.00469   Mean   : 0.0979   Mean   :0.2018
##  3rd Qu.: 0.04048   3rd Qu.: 0.1402   3rd Qu.:0.2438
##  Max.   : 0.21542   Max.   : 0.3356   Max.   :0.4084
##      b0.lm             b1.lm              b2.lm
##  Min.   :0.456   Min.   :-0.0317   Min.   :0.00493
##  1st Qu.:0.487   1st Qu.: 0.0135   1st Qu.:0.03939
##  Median :0.499   Median : 0.0241   Median :0.04979
##  Mean   :0.499   Mean   : 0.0241   Mean   :0.04959
##  3rd Qu.:0.510   3rd Qu.: 0.0346   3rd Qu.:0.05988
##  Max.   :0.553   Max.   : 0.0811   Max.   :0.09825
```

It can be seen that we have produced a matrix with 1,000 rows from the 1,000 simulations and 6 columns with 3 columns each for the three regression parameters form the *logistic regression model* and *linear regression model.* With these 1,000 simulations, we can calculate the sampling means from these 1,000 samples using *R* function *apply* as follows:

```
apply(estParMat,2,mean)
```

```
##     b0.glm    b1.glm    b2.glm     b0.lm     b1.lm    b2.lm
## -0.004685  0.097859  0.201761  0.498844  0.024099 0.049592
```

This shows that the overall sampling means for $(\hat{\beta}_0, \hat{\beta}_1, \hat{\beta}_2)$ are (-0.004685422, 0.097858940, 0.201760902) from the *logistic regression* and (0.498844000, 0.024099194, 0.049591809) from the *linear regression*, respectively, where the true parameters are $(\beta_0, \beta_1, \beta_2) = (0, 0.1, 0.2)$. This is evident that the estimated regression parameters for the *linear regression* are biased, but they are unbiased with the *logistic regression model.* This evidence can be graphically shown in the following Figure 5.3, which can be produced by the following *R* code chunk:

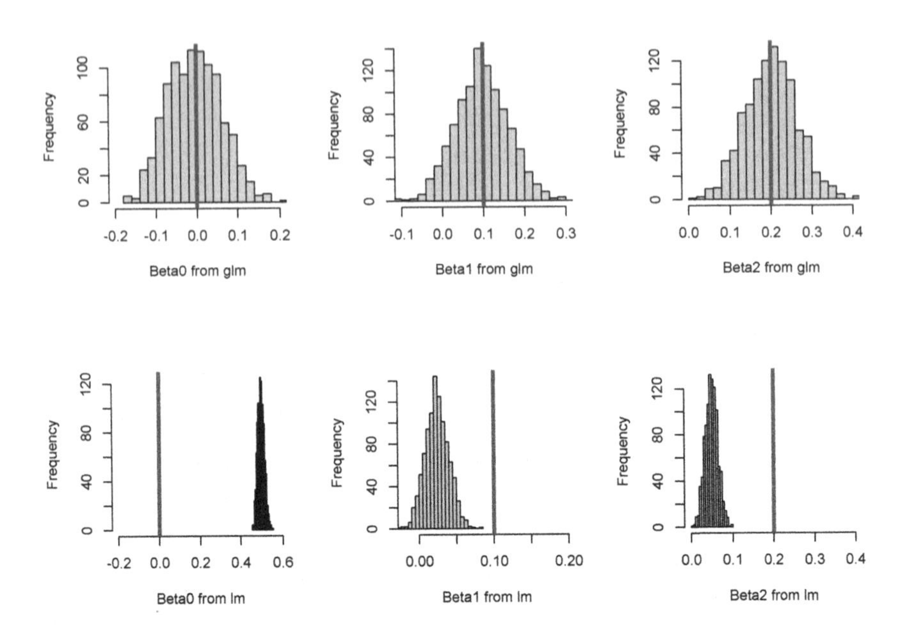

FIGURE 5.3
Monte-Carlo Distributions Overlaid with the True Parameters

```
# Figure setup
par(mfrow=c(2,3))
# Make the plots for logistic regression model
hist(estParMat[,1],xlim=c(-0.2,0.2),
    nclass=20, main="", xlab="Beta0 from glm")
abline(v=0, lwd=3, col="red")
hist(estParMat[,2],xlim=c(-0.1,0.3),
    nclass=20,main="", xlab="Beta1 from glm")
abline(v=0.1, lwd=3, col="red")
hist(estParMat[,3],xlim=c(0,0.4),
    nclass=20,main="", xlab="Beta2 from glm")
abline(v=0.2, lwd=3, col="red")
# Make the plots for linear regression model
hist(estParMat[,4], xlim=c(-0.2,0.6),
    nclass=20, main="", xlab="Beta0 from lm")
abline(v=0, lwd=3, col="red")
hist(estParMat[,5], xlim=c(-0.02,0.2),
    nclass=20,  main="", xlab="Beta1 from lm")
abline(v=0.1, lwd=3, col="red")
hist(estParMat[,6],xlim=c(0,0.4),
```

```
    nclass=20,  main="", xlab="Beta2 from lm")
abline(v=0.2, lwd=3, col="red")
```

Observing the plots presented in the upper panel of Figure 5.3, it becomes apparent that the three parameters estimated via the *logistic regression model* exhibit an absence of bias. These estimated parameters align closely with the true values, as evidenced by the congruence with the *red* vertical lines denoting ground truth. In contrast, when we examine the plots in the lower panel of Figure 5.3, an opposing pattern emerges. Here, the three parameters that fitted by the *linear regression model* display evidence of bias, as discernible from the visual discrepancy relative to the expected true values displayed by the *red* vertical lines.

5.4 Discussions

In this chapter, we explored the shift from analyzing normally distributed continuous data to binary/binomial outcomes within the realm of financial data analysis. Logistic regression emerged as a powerful statistical technique to handle these binary outcomes effectively. We used the example of S&P stock market data from 2001 to 2005 to demonstrate the practical application of these analytical methods, emphasizing their implementation in the *R* programming language.

The logistic regression model serves as a valuable tool to fit and predict binary outcomes in financial statistics. By constructing models that estimate the probability of stock success based on various independent variables related to the stock market, we can make predictions about whether the market will experience a rise or fall.

This approach enables us to leverage historical data to uncover relationships between market variables and binary outcomes in financial market. Logistic regression's ability to model probabilities provides insights into the likelihood of specific events occurring, such as stock market movements, enabling investors and analysts to make informed decisions and manage risks effectively.

As with any analytical method, it's important to exercise caution and consider the assumptions and limitations of logistic regression. The quality of predictions hinges on the quality and relevance of the independent variables used in the model, and a comprehensive understanding of financial market dynamics is crucial for accurate interpretation.

Overall, logistic regression stands as a valuable tool for addressing binary/binomial outcomes in financial data analysis, enabling professionals to enhance decision-making and gain deeper insights into market behaviors.

5.5 Exercises

It is known that there is a significant correlation between the hourly earnings (denoted by *Earnings*) and education levels (denoted by *EDU*), which can be described by the regression equation:

$$\begin{aligned} Y(Earnings) &= \beta_0 + \beta_1 \times EDU + \epsilon \\ &= 5.40 + 1.06 \times EDU + \epsilon \end{aligned}$$

where the error term $\epsilon \sim N(0, \sigma^2)$ with $\sigma = 1.5$. In this regression line, the coefficient β_1 tells us that an additional year of education is associated with a *1.06* increase in the hourly wage. And for those with no education ($EDU = 0$), the intercept β_0 indicates that the average wage is *5.40* an hour.

1. Based on the above fact, simulate $n = 1{,}000$ study participants with *EDU* between 5 (*elementary school education*) and 20 (*Graduate school education*).

   Hint: Using the following *R* code chunk:

```
# Set random seed
set.seed(333)
# Number of participants
n = 1000
# Random sample for "EDC"
EDU = sample(seq(5,20), n, replace=TRUE)
```

2. Calculate their corresponding *Earnings* using the regression equation $Y(Earnings) = 5.40 + 1.06 \times EDU + \epsilon$ and plot the relationship between the education levels and hourly earnings.

   Hint: Using the following *R* code chunk:

```
# The error variance
sigma = 1.5
# Generate error term
err = rnorm(n, 0, sigma^2)
# Calculate the "Earnings"
Y = 5.40 + 1.06*EDU + err
# Plot the relationship
plot(EDU, Y, xlab="Education Levels",
    ylab="Hourly Earnings")
```

3. Fit the linear regression $Y = \beta_0 + \beta_1 \times EDU + \epsilon$ to the data and show the estimated regression parameters. Answer the following questions:

   - Is the regression model significant?
   - Are the estimated regression parameters $(\hat{\beta}_0, \hat{\beta}_1)$ close to the true parameters $(\beta_0, \beta_1) = (5.40, 1.06)$?
   - Is the estimated error standard deviation $\hat{\sigma}$ close to the true value $\sigma = 1.5$?

   Hint: Using the following *R* code chunk:

```
# Fit linear regression using lm
lm.Edc2Earnings = lm(Y~EDC)
# Print the fit
summary(lm.Edc2Earnings)
```

4. Dichotomize the *Earnings* data Y at the average earnings (denoted by $m.Y$) to create a new variable $catY$ where $catY = 1$ if $Y > m.Y$, otherwise $catY = 0$.

   Hint: using the following *R* code:

```
# Get the mean of "Earnings"
m.Y = mean(Y)
# Create a binary variable on
catY = rep(0,n); catY[Y >m.Y] =1
```

5. Perform a *logistic regression* to regress the $catY$ to EDU and answer the following questions:

   - Is this logistic regression significant?
   - What are the estimated parameters?
   - Are these parameter estimates close to the true parameters of $(\beta_0, \beta_1) = (5.40, 1.06)$? If not, why?

6

The Poisson Regression: Models for Count Data

Another important and commonly seen data type in financial data analysis is count (also known as counts) data, where the variable we will consider represents the number of events that occur within a fixed interval of time or space. Examples of these include the number of trades executed in the stock markets, the number of defaults in insurance industries, or the number of market events within a specific time period.

This type of data is different from the binomial data we considered in Chapter 5. Even binomial data and count data are both types of discrete data, binomial data refers to data that arise from a binomial experiment which is characterized by the total success trials among a pre-fixed number of trials (n_i) and known in advance whereas the count data is characterized by just the total numbers without the pre-specified number of trials. Examples of count data in finance could include the number of trades executed in a given hour (not in a given total number of trades), the number of customer transactions in a day, or the number of stock price changes during a trading session.

Because of this difference, binomial data often follows a binomial distribution, and analysis of such data involve methods of logistic regression as presented in Chapter 5. Count data often follows a Poisson or negative-binomial distribution. Analysis would involve methods like Poisson regression or negative binomial regression, which is the focus of this chapter. Even the negative-binomial regression is not part of the *GLM*, but the Poisson regression is part of the *GLM* which has been widely introduced and well explained. Interested readers can refer to McCullagh and Nelder (1995), Dobson and Barnett (2018), and Dunn and Smyth (2018). In this chapter, we will briefly describe the theoretical background of Poisson regression and the associated extension to model underdispersion and overdispersion. We do this by using the quasi-Poisson regression with the *R* function *glm*, and the negative-binomial regression with the *R* function *glm-nb*.

We will use two datasets in this chapter to illustrate the modeling of count data. The first dataset is the credit card data *CreditCard*, which contains credit history for a sample of applicants for a type of credit card. This dataset is one of the datasets from the *R* library *AER*, which includes all the functions,

DOI: 10.1201/9781003469704-6

data sets, examples for the book *Applied Econometrics with R* by Kleiber and Zeileis (2008). We will model the outcome variable *active* (Number of active credit card accounts) to other variables in this dataset since the number of active credit accounts should be modeled as a Poisson model. This variable also exhibits large *overdispersion*, so the quasi-Poisson regression and negative-binomial regression should be more appropriate. The second dataset is the *Smarket* data from Chapter 5, used in the previous chapter. We will model *Volume*, i.e., the volume of shares traded. This is the count of total daily shares traded in billions using the percentage returns from the previous 5 days (*Lag1* to *Lag5*) since. This data is an example of typical *underdispersion* count data.

Additionally, we will design and conduct a Monte-Carlo simulation study to investigate the effects of *dispersion* on the conventional *Poisson regression*. In this investigation, we'll examine how *overdispersion* can lead to Type-I errors by producing an excessive number of false-positive significances. Conversely, we'll explore how *underdispersion* can lead to Type-II errors, characterized by an excessive prevalence of false-negative significances. This comprehensive analysis holds particular relevance within the domain of financial data analysis.

6.1 Why Poisson Regression to Model Count Data

Let's describe the data first and we can then answer the question of why we need to have a new model (i.e., the Poisson regression) to model the count data.

6.1.1 Data Description

The *CreditCard* data is a small credit card dataset for simple econometric analysis from the *R* package *AER* with 1,319 observations on 12 variables as follows:

- *card*: a factor variable to answer the question: Was the application for a credit card accepted? 1 if application for credit card is accepted, 0 if not,
- *reports*: the number of major derogatory reports,
- *age*: age in years plus twelfths of a year,
- *income*: yearly income in US dollars (divided by 10,000),
- *share*: ratio of monthly credit card expenditure to yearly income,
- *expenditure*: average monthly credit card expenditure,
- *owner*: a factor variable to answer the question: Does the individual own their home? 1 if owns their home, 0 if rent,

- *selfempl*: a factor variable to answer the question: Is the individual self-employed? 1 if self-employed, 0 if not,
- *dependents*: number of dependents,
- *months*: months living at current address,
- *majorcards*: number of major credit cards held,
- *active*: number of active credit accounts.

This data can be loaded to the *R* session and we can see the summary of the data:

```
# Load the AER library
library(AER)

## Loading required package: sandwich

## Loading required package: survival

# Load the data and print the summary
data(CreditCard); summary(CreditCard)

##   card          reports             age
##  no : 296   Min.    : 0.000   Min.    : 0.17
##  yes:1023   1st Qu.: 0.000   1st Qu.:25.42
##              Median : 0.000   Median :31.25
##              Mean   : 0.456   Mean   :33.21
##              3rd Qu.: 0.000   3rd Qu.:39.42
##              Max.   :14.000   Max.   :83.50
##      income           share           expenditure
##  Min.   : 0.21   Min.   :0.0001   Min.   :    0.0
##  1st Qu.: 2.24   1st Qu.:0.0023   1st Qu.:    4.6
##  Median : 2.90   Median :0.0388   Median : 101.3
##  Mean   : 3.37   Mean   :0.0687   Mean   : 185.1
##  3rd Qu.: 4.00   3rd Qu.:0.0936   3rd Qu.: 249.0
##  Max.   :13.50   Max.   :0.9063   Max.   :3099.5
##  owner     selfemp      dependents         months
##  no :738   no :1228   Min.   :0.000   Min.   :  0.0
##  yes:581   yes:  91   1st Qu.:0.000   1st Qu.: 12.0
##                        Median :1.000   Median : 30.0
##                        Mean   :0.994   Mean   : 55.3
##                        3rd Qu.:2.000   3rd Qu.: 72.0
##                        Max.   :6.000   Max.   :540.0
##    majorcards         active
##  Min.   :0.000   Min.   : 0
```

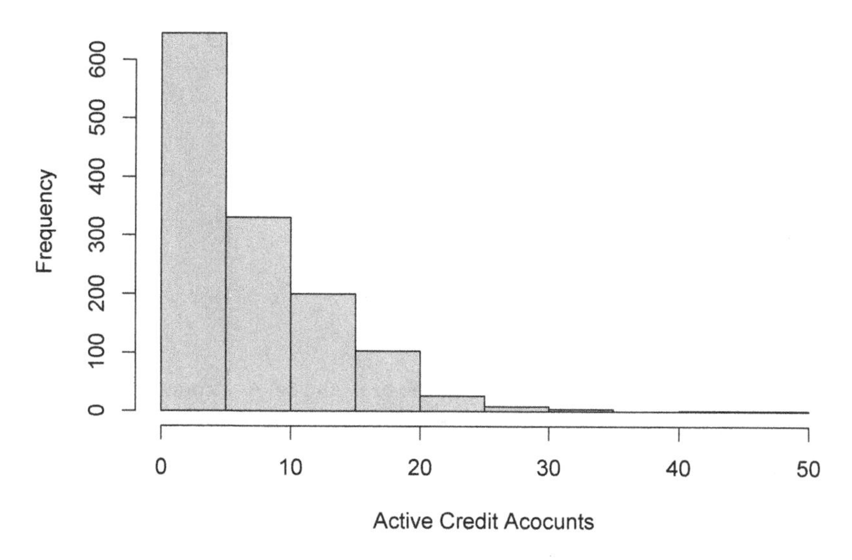

FIGURE 6.1
Distribution of the Number of Active Credit Cards

```
##  1st Qu.:1.000   1st Qu.: 2
##  Median :1.000   Median : 6
##  Mean   :0.817   Mean   : 7
##  3rd Qu.:1.000   3rd Qu.:11
##  Max.   :1.000   Max.   :46
```

6.1.2 Why Not Linear Regression

The logical question for when you have count data is, why can linear regression not be used anymore? There are many reasons to be discussed with the major reason being that count data are not normally distributed anymore. Count data are inherently discrete and nonnegative, and they often follow a skewed distribution. We can take a look at the credit card data *CreditCard* and plot the data distribution as follows:

```
# Plot the distribution
hist(CreditCard$active,
     xlab="Active Credit Acocunts", main="")
```

It is obvious that this distribution is not normally distributed as shown in Figure 6.1. It is in fact heavily right-skewed since there are fewer and fewer people who have a large number of credit cards.

This can be further investigated if we forcefully fit a linear regression as follows:

```
# Fit the linear regression
modCC.lm =lm(active~age+income+share+expenditure
             +owner+selfemp+dependents+months,
             data = CreditCard)
# Print the model fit
summary(modCC.lm)
```

```
##
## Call:
## lm(formula = active ~ age + income + share + expenditure +
## owner + selfemp + dependents + months, data = CreditCard)
##
## Residuals:
##    Min      1Q Median      3Q     Max
## -13.26  -4.51  -1.21   3.40  36.32
##
## Coefficients:
##              Estimate Std. Error t value Pr(>|t|)
## (Intercept)  3.599517   0.681797    5.28  1.5e-07 ***
## age          0.041864   0.019808    2.11    0.035 *
## income       0.223622   0.131564    1.70    0.089 .
## share       -5.130233   3.902137   -1.31    0.189
## expenditure  0.001887   0.001408    1.34    0.180
## owneryes     2.813823   0.380883    7.39  2.7e-13 ***
## selfempyes   0.083432   0.662858    0.13    0.900
## dependents  -0.029914   0.145374   -0.21    0.837
## months       0.000768   0.002816    0.27    0.785
##
## Residual standard error: 6.03 on 1310 degrees of freedom
## Multiple R-squared:  0.0905, Adjusted R-squared:  0.0849
## F-statistic: 16.3 on 8 and 1310 DF,  p-value: <2e-16
```

It seems that we have found a significant regression, but we can check the residual distribution using *QQ-plot* as follows:

```
# Call qqnorm to plot the residual quantiles
qqnorm(modCC.lm$residuals)
# Add a line to the QQplot
qqline(modCC.lm$residuals, lwd=3, col="red")
```

We can see a large deviation from the normal distribution in Figure 6.2, which is indeed the case since the data are counts. Therefore, the classical linear regression is not appropriate to model counts data and a new regression model

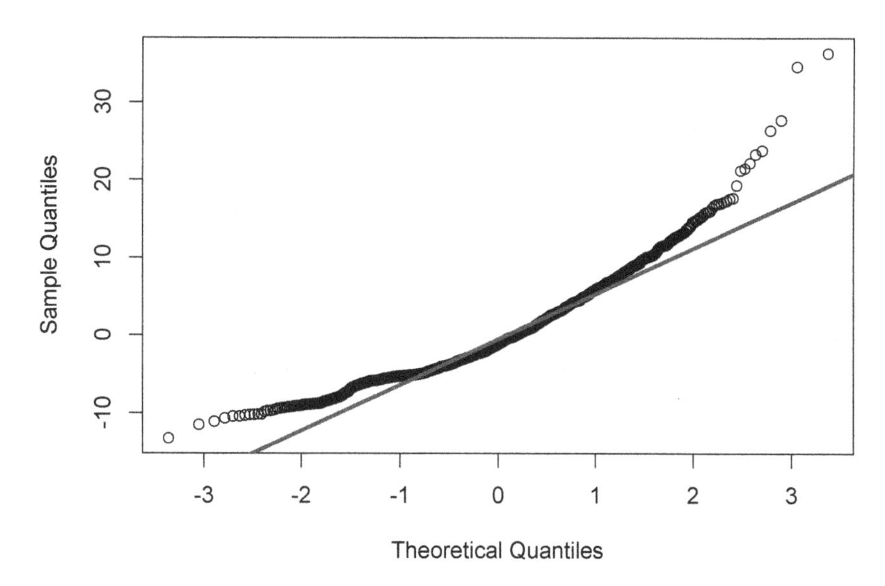

FIGURE 6.2
QQ-Plot of the Residual from Linear Regression

should be developed. This new model is the Poisson regression model. Poisson regression is a statistical technique specifically designed for modeling count data, making it a suitable choice when you're dealing with situations where the dependent variable represents the number of occurrences of an event.

6.1.3 Why Not the Log-Linear Regression

Another classic practice is to use a log-transformation on count data, since the log-transformation is commonly used to stabilize variance and approximate normality. However, count data often includes zeros, and the logarithm of zero is undefined. This can lead to problems when attempting to log-transform data that contains zeros and would result in *undefined values* or *missing values.*

This is the exact case in this *CreditCard* data where there is a large proportion of customers with zero *active* credit cards. This can be illustrated using the following *R* code chunk:

```
# Count the number of customers with zero active credit cards
num0 = sum(CreditCard$active==0); num0
```

```
## [1] 219
```

```
# Count the total number of customers
num.tot = dim(CreditCard)[1];num.tot
```

```
## [1] 1319
```

```
# Calculate the proportion
num0/num.tot
```

```
## [1] 0.166
```

Therefore among the 1,319 customers, there are 219 customers who have zero active credit cards, which is 16.6% of the total customers. If we take a log-transformation of these 0's, we would produce 217 undefined values as denoted by *-Inf* in *R* and seen in the following *R* code chunk:

```
# Log-transform the count data
log_active = log(CreditCard$active)
# Print the undefined values
log_active[log_active == -Inf]
```

```
##   [1] -Inf -Inf -Inf -Inf -Inf -Inf -Inf -Inf -Inf -Inf
##  [11] -Inf -Inf -Inf -Inf -Inf -Inf -Inf -Inf -Inf -Inf
##  [21] -Inf -Inf -Inf -Inf -Inf -Inf -Inf -Inf -Inf -Inf
##  [31] -Inf -Inf -Inf -Inf -Inf -Inf -Inf -Inf -Inf -Inf
##  [41] -Inf -Inf -Inf -Inf -Inf -Inf -Inf -Inf -Inf -Inf
##  ...(deleted)....
## [141] -Inf -Inf -Inf -Inf -Inf -Inf -Inf -Inf -Inf -Inf
## [151] -Inf -Inf -Inf -Inf -Inf -Inf -Inf -Inf -Inf -Inf
## [161] -Inf -Inf -Inf -Inf -Inf -Inf -Inf -Inf -Inf -Inf
## [171] -Inf -Inf -Inf -Inf -Inf -Inf -Inf -Inf -Inf -Inf
## [181] -Inf -Inf -Inf -Inf -Inf -Inf -Inf -Inf -Inf -Inf
## [191] -Inf -Inf -Inf -Inf -Inf -Inf -Inf -Inf -Inf -Inf
## [201] -Inf -Inf -Inf -Inf -Inf -Inf -Inf -Inf -Inf -Inf
## [211] -Inf -Inf -Inf -Inf -Inf -Inf -Inf -Inf -Inf
```

If we use the log-transformed data for further analysis, the sample size would reduce from 1,319 to 1,100 (= 1319-219). This can be further investigated if we forcefully fit a linear regression as follows:

```
# Include "log_active" into the "CreditCard"
CreditCard$log_active = log(CreditCard$active)
# Fit the loglinear regression to the
modCC.loglm =lm(log_active~age+income+share+expenditure
```

```
              +owner+selfemp+dependents+months,
              data = CreditCard[CreditCard$log_active !=-Inf,])
# Print the model fit
summary(modCC.loglm)
```

```
##
## Call:
## lm(formula = log_active ~ age + income + share + expenditure +
##     owner + selfemp + dependents + months,
##     data = CreditCard[CreditCard$log_active != -Inf, ])
##
## Residuals:
##    Min     1Q Median     3Q    Max
## -2.248 -0.492  0.091  0.556  1.882
##
## Coefficients:
##              Estimate Std. Error t value Pr(>|t|)
## (Intercept)  1.354864   0.096709   14.01  < 2e-16 ***
## age          0.005982   0.002761    2.17    0.030 *
## income       0.048655   0.019231    2.53    0.012 *
## share       -1.044606   0.569630   -1.83    0.067 .
## expenditure  0.000309   0.000205    1.51    0.132
## owneryes     0.382950   0.052852    7.25  8.1e-13 ***
## selfempyes  -0.182344   0.089577   -2.04    0.042 *
## dependents   0.018285   0.020796    0.88    0.379
## months      -0.000832   0.000386   -2.15    0.032 *
##
## Residual standard error: 0.777 on 1091 degrees of freedom
## Multiple R-squared:  0.118,  Adjusted R-squared:  0.111
## F-statistic: 18.2 on 8 and 1091 DF,  p-value: <2e-16
```

Different statistical estimates and statistical significance can be observed if comparing this log-transformed linear regression model in *modCC.loglm* with the original linear regression in *modCC.lm*. This is not surprising since these two models are different and furthermore the number of observations (1,319 in *modCC.lm*, vs. 1,100 in *modCC.loglm*) are also different due to the log-transformation.

As mentioned, the log-transformation is commonly used to stabilize variance and approximate normality. We can verify this detail with the fitted log-linear model *modCC.loglm* to examine the residuals as follows:

```
# Figure layout
par(mfrow=c(1,2))
```

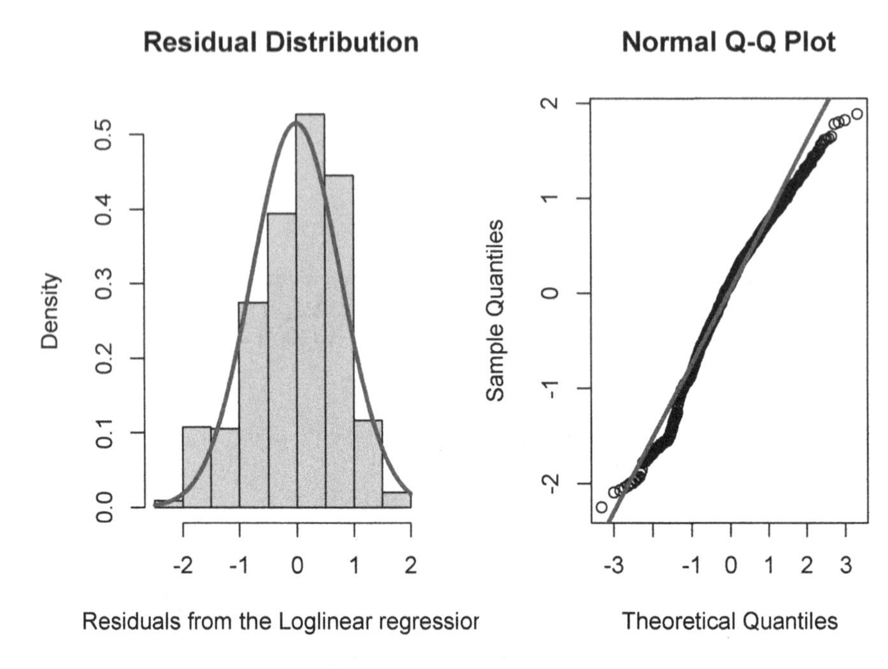

FIGURE 6.3
Residual Distribution for the Log-Transformed Linear Regression Model

```
# Extract the residuals
resid.loglm = modCC.loglm$residuals
# Plot the residual distribution
hist(resid.loglm, prob=T,
     xlab="Residuals from the Loglinear regression",
     main="Residual Distribution")
# Overlay the normal distribution on the histogram
dn = function(x)
  dnorm(x,mean=mean(resid.loglm), sd=sd(resid.loglm))
curve(dn, col="red", lwd=3, add=TRUE)
# Call qqnorm to plot the residual quantiles
qqnorm(resid.loglm)
# Add a line to the qqplot
qqline(resid.loglm, lwd=3, col="red")
```

As seen in Figure 6.3, the residual distribution in the left panel seems more normally distributed. The associated QQ-plot in the right panel also looks much better in line with the 1:1 line comparing with the QQ-plot in Figure 6.2. However, if a formal statistical test is used (i.e., the *Shapiro test for normality*), we still reject the hypothesis that the residual distribution is normally distributed as seen below:

```
## Shapiro test for normality
shapiro.test(resid.loglm)
```

```
##
##  Shapiro-Wilk normality test
##
## data:  resid.loglm
## W = 0.98, p-value = 3e-11
```

When dealing with count data and the challenges associated with log-transforming it, as demonstrated above, there's another common practice of adding a small constant (such as 1) to the count values before performing the log-transformation. This helps to avoid undefined values that would arise from attempting to take the logarithm of zero. However, this approach introduces a new question: why choose 1 and not another value, such as 2 or any other constant? In fact, there isn't a universally optimal constant to add, and the choice can be somewhat arbitrary. Even adding a constant of 1 is a common choice to ensure that all zero counts are shifted to non-zero values before transformation, the resulted distributions are totally different and this dilemma can be easily shown using the following *R* code chunk:

```
# Add a small value 1
CreditCard$log_active1 = log(CreditCard$active+1)
# Histograms
par(mfrow=c(1,2))
hist(CreditCard$log_active,
     xlab="Log-Transformed",main="")
hist(CreditCard$log_active1,
     xlab="Log-Transformed with 1 Added", main="")
```

As observed from the right panel in Figure 6.4, the histogram plot now incorporates a prominent bar at the left representing the 219 zero values. This inclusion noticeably distorts the distribution, causing it to deviate further from the normal distribution.

In summary, the decision to add a constant and the choice of the constant itself should be made with careful consideration of the data's characteristics and the goals of the analysis. While this approach can mitigate some issues associated with log-transforming count data, it's important to understand that it doesn't fully address all the complexities tied to count data distributions and can bring more questions to the analysis. Therefore, alternative statistical methods

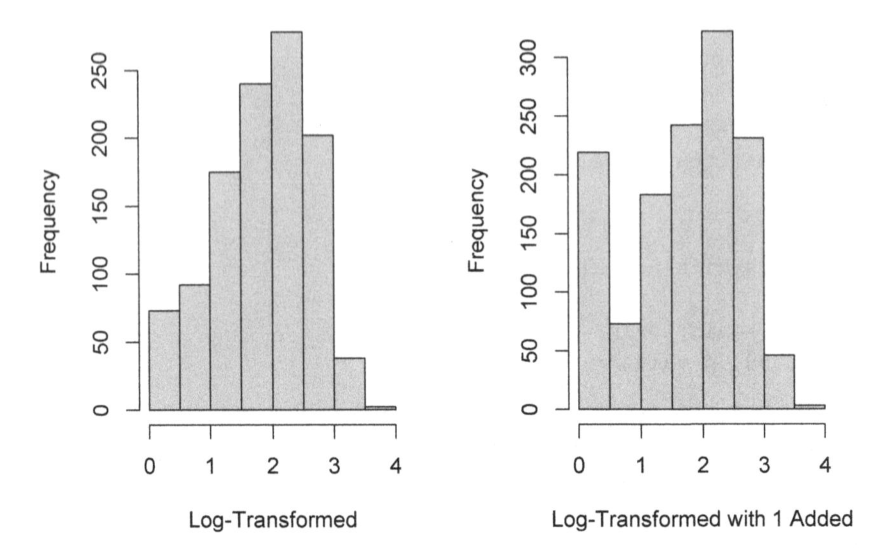

FIGURE 6.4
Distributions between Log-Transformed with/without 1 Added

designed for count data are needed, which is again the *Poisson regression* in the *Generalized Linear Models (GLMs).*

6.2 Poisson Regression

6.2.1 Conventional Poisson Regression

Poisson regression was developed as part of the *generalized linear model (GLM)* as discussed in McCullagh and Nelder (1995) to model count data.

For a count variable Y to be Poisson distributed, its probability distribution is written as:

$$P(Y = y) = \frac{\mu^y e^{\mu}}{y!} \tag{6.1}$$

where $y = 0, 1, 2, \cdots$. In this definition in equation (6.1), both the mean and variance of Y are μ.

In this *generalized linear model framework*, we use *a linear predictor* to model the linear relationship and a *link function* to link the *linear predictor* to the count data y. Similarly to the logistic regression, the *linear predictor* is denoted by:

$$\eta_i = \beta_0 + \beta_1 x_{i1} + \cdots + \beta_K x_{iK} = X_i \beta \tag{6.2}$$

where $X_i = (1, x_{i1}, \cdots, x_{iK})$ is the matrix of observed covariates and $\beta = (\beta_0, \beta_1, \cdots, \beta_K)$ is the associated parameter vector.

To make sure the counts are modeled and predicted as non-negative values, a log *link function* is used by default in Poisson regression to overcome the problem of negative predicted counts, since the log of the counts can take any real numbered value as follows:

$$log(\mu) = \eta = \beta_0 + \beta_1 x_1 + \cdots + \beta_K x_K, \tag{6.3}$$

or equivalently

$$\mu = exp(\eta) = exp\left(\beta_0 + \beta_1 x_1 + \cdots + \beta_K x_K\right). \tag{6.4}$$

With this log link function in equation (6.4), we can then model the mean μ from the count data Y to all the covariates Xs using the maximum likelihood estimation. We will focus on *R* implementation with *glm* function and skip the mathematical derivations and interested readers can refer to McCullagh and Nelder (1995).

Poisson regression is part of the *R* function *glm* and it is very straight to fit the Poisson regression as follows:

```
# Fit the Poisson regression
modCC.poisson = glm(active~age+income+share+expenditure
             +owner+selfemp+dependents+months,CreditCard,
             family =c("poisson"))
# Print the model fit
summary(modCC.poisson)
```

```
##
## Call:
## glm(formula = active ~ age + income + share + expenditure
## + owner + selfemp + dependents + months,
## family = c("poisson"), data = CreditCard)
##
## Coefficients:
##              Estimate Std. Error z value Pr(>|z|)
## (Intercept)  1.45e+00   4.28e-02   33.95  < 2e-16 ***
## age          5.82e-03   1.21e-03    4.81  1.5e-06 ***
## income       2.65e-02   7.44e-03    3.56  0.00038 ***
## share       -6.37e-01   2.35e-01   -2.71  0.00672 **
## expenditure  2.07e-04   7.38e-05    2.80  0.00511 **
```

```
## owneryes      4.06e-01   2.38e-02   17.03  < 2e-16 ***
## selfempyes    1.32e-02   3.98e-02    0.33  0.74022
## dependents   -1.74e-03   8.72e-03   -0.20  0.84191
## months        3.93e-05   1.66e-04    0.24  0.81280
##
## (Dispersion parameter for poisson family taken to be 1)
##
##     Null deviance: 7795  on 1318  degrees of freedom
## Residual deviance: 7135  on 1310  degrees of freedom
## AIC: 11239
##
## Number of Fisher Scoring iterations: 5
```

As seen from the above *R* code chunk, we used probability distribution *family =c("poisson")* in *glm* to fit the Poisson regression. The result showed that the *age*, *income*, *share*, *expenditure*, and *owner* are significant. However, before we make any conclusion, we need to examine whether or not this model fits the data by examining the residual deviance. As discussed in the logistic regression, the residual deviation is χ^2 distributed and can be used to test the null hypothesis that the model fits the data. We extract the residual deviance and the associated degrees of freedom for this calculation as follows:

```
# p-value
pval.poisson= 1-pchisq(deviance(modCC.poisson),
                       df.residual(modCC.poisson))
# print
pval.poisson
```

```
## [1] 0
```

The resulting p-value < 0.0001, which is highly significant, suggesting that the conventional *Poisson regression model modCC.poisson* did not fit the data adequately and that further exploration is needed. This lack of goodness-of-fit is probably due to the data collection where there are other important predictive variables missing and they are correlated with the number of active credit cards.

6.2.2 Models for Overdispersed Count Data

The fundamental assumption in Poisson regression is that the counts are assumed to be Poisson distributed and the underlying assumption of Poisson distribution is that the mean is equal to the variance, i.e. $\sigma^2 = var(Y) = mean(Y) = \mu$, as seen in equation (6.1).

To verify whether this is the case for the credit card data *CreditCard*, we can easily calculate them as follows:

```
# Calculate the variance and mean
var.CC  = var(CreditCard$active); var.CC

## [1] 39.76

mean.CC = mean(CreditCard$active); mean.CC

## [1] 6.997

# the ratio between the variance and mean
var.CC/mean.CC

## [1] 5.683
```

From this calculation, we can see that the variance is 39.763 and the mean is 6.997. The variance is 5.683 times larger than the mean, indicating that the fundamental assumption of Poisson distribution is violated. This phenomenon is called overdispersion in counts regression.

Similarly, we can verify the mean and variance relationship for the *Smarket* data as follows:

```
# load the library which includes the data
library(ISLR2)
# load the data into R and check the dimension
data(Smarket);
# Calculate the variance and mean
var.Smarket   = var(Smarket$Volume); var.Smarket

## [1] 0.1299

mean.Smarket = mean(Smarket$Volume); mean.Smarket

## [1] 1.478

# the ratio between the variance and mean
var.Smarket/mean.Smarket

## [1] 0.08784
```

For the *Smarket* data, we found an underdispersion since the variance is 0.1298 and the mean is 1.4783. The variance is only 0.0878 times the mean, indicating again that the fundamental assumption of Poisson distribution is violated. This phenomenon is called underdispersion in counts regression.

To deal with overdispersion, two approaches are usually used with one called *Quasi-Poisson regression* and another called *negative-binomial regression*. However, only the *Quasi-Poisson regression* can be used to deal with underdispersion.

6.2.2.1 Quasi-Poisson Regression

In quasi-Poisson regression, the overdispersion or underdispersion is incorporated with the dispersion parameter ϕ, i.e., $\sigma^2 = var(Y) = \phi \times mean(Y) = \phi\mu$. Therefore when ϕ is greater than 1, we can model the overdispersion and when ϕ is less than 1, we can model the underdispersion. The classical Poisson regression is then the special case that $\phi = 1$. Therefore, the quasi-Poisson regression is an extension of the classical Poisson regression. This overdispersion or the underdispersion parameter will be estimated along with all the βs from the maximum likelihood estimation.

In fitting the quasi-Poisson regression, the overdispersion parameter is estimated based on the data and then used to adjust the standard errors for all the estimated parameters. Therefore, the estimated values of the parameters βs will not change, but the estimated standard error, the associated t-values and p-values will be adjusted and updated from the conventional *Poisson regression*. The *quasi-Poisson regression* can be implemented in *glm* as follows:

```
modCC.quasipoisson = glm(active~age+income+share+expenditure
            +owner+selfemp+dependents+months,CreditCard,
            family = c("quasipoisson"))
# summary
summary(modCC.quasipoisson)
```

```
##
## Call:
## glm(formula = active ~ age + income + share + expenditure +
## owner + selfemp + dependents + months,
## family = c("quasipoisson"), data = CreditCard)
##
## Coefficients:
##              Estimate Std. Error t value Pr(>|t|)
## (Intercept)  1.45e+00   9.81e-02   14.82  < 2e-16 ***
## age          5.82e-03   2.77e-03    2.10    0.036 *
## income       2.65e-02   1.71e-02    1.55    0.121
## share       -6.37e-01   5.38e-01   -1.18    0.237
## expenditure  2.07e-04   1.69e-04    1.22    0.222
## owneryes     4.06e-01   5.46e-02    7.43  1.9e-13 ***
## selfempyes   1.32e-02   9.13e-02    0.14    0.885
## dependents  -1.74e-03   2.00e-02   -0.09    0.931
```

```
## months        3.93e-05   3.80e-04    0.10    0.918
##
## (Dispersion parameter for quasipoisson family
## taken to be 5.249)
##
##     Null deviance: 7795  on 1318  degrees of freedom
## Residual deviance: 7135  on 1310  degrees of freedom
## AIC: NA
##
## Number of Fisher Scoring iterations: 5
```

As seen from the output, the dispersion parameter is estimated at 5.249459 as seen *(Dispersion parameter for quasipoisson family taken to be 5.249359)* in the *modCC.quasipossion* model where the dispersion parameter is defaulted to 1 in the classical Poisson regression *modCC.poisson* as seen *(Dispersion parameter for Poisson family taken to be 1)*. With the estimated dispersion parameter of 5.249459, we can conclude that there is an overdispersion in the variable of the number of active credit cards. Adjusting this estimated dispersion parameter, the standard errors of all the parameters are scaled by 5.249459, which is larger than those in the classical Poisson regression *modCC.poisson*. With this adjustment, the statistical significance for all parameters is reduced. Noticeably the only *age* and *owner* are statistically significant in the quasi-Poisson regression *modCC.quasipoisson*, whereas the *age*, *income*, *share*, *expenditure*, and *owner* are significant in the conventional *Poisson regression model* in *modCC.poisson*.

6.2.2.2 Negative-Binomial Regression

When dealing with overdispersed count data where the variance exceeds the mean, negative-binomial regression is a commonly used alternative to the classical Poisson regression. The negative-binomial regression model is based on the negative binomial distribution, which is a generalization of the Poisson distribution. In the negative-binomial distribution, the mean is identical to that of the Poisson; however, the variance is

$$Var(Y) = \mu + \frac{\mu^2}{\theta} \tag{6.5}$$

Therefore in negative-binomial distribution, an extra term $\frac{\mu^2}{\theta}$ is added to the mean μ. In this model, it is clear that as θ increases and approaches infinity, the variance approaches the mean. In this case, the negative binomial distribution becomes more like the classical Poisson distribution. When $\theta = 1$, this negative-binomial distribution becomes the Gamma distribution. In this special case, the negative-binomial regression will be the *Gamma regression*.

The negative-binomial regression is implemented in *R* function *glm.nb* within the *MASS* library (i.e., functions and datasets to support Venables and Ripley,

Modern Applied Statistics with S). For this example, the implementation to fit the negative-binomial regression is as follows:

```
# Load the MASS library
library(MASS)
# fit negative binomial model
modCC.nb = glm.nb(active~age+income+share+expenditure
              +owner+selfemp+dependents+months,
              data = CreditCard)
# summary
summary(modCC.nb)
```

```
##
## Call:
## glm.nb(formula = active ~ age + income + share + expenditure +
##     owner + selfemp + dependents + months, data = CreditCard,
##     init.theta = 1.196062143, link = log)
##
## Coefficients:
##              Estimate Std. Error z value Pr(>|z|)
## (Intercept)  1.397368   0.112351   12.44  < 2e-16 ***
## age          0.006906   0.003249    2.13    0.034 *
## income       0.032477   0.021359    1.52    0.128
## share       -0.603935   0.641742   -0.94    0.347
## expenditure  0.000210   0.000228    0.92    0.355
## owneryes     0.400651   0.062494    6.41  1.4e-10 ***
## selfempyes  -0.010281   0.108658   -0.09    0.925
## dependents  -0.010077   0.023810   -0.42    0.672
## months       0.000180   0.000459    0.39    0.696
## ---
##
## (Dispersion parameter for Negative Binomial(1.196)
## family taken to be 1)
##
##     Null deviance: 1671.1  on 1318  degrees of freedom
## Residual deviance: 1573.1  on 1310  degrees of freedom
## AIC: 7874
##
## Number of Fisher Scoring iterations: 1
##
##
##               Theta:  1.1961
##           Std. Err.:  0.0606
##
##  2 x log likelihood:  -7853.8240
```

As shown above, the estimated overdispersion parameter is $\hat{\theta} = 1.1961$ with a standard error of 0.0606. Similar statistical significance can be observed in the negative-binomial regression to the quasi-Poisson regression, where only *age* and *owner* are statistically significant in comparison with the conventional *Poisson regression* in *modCC.poisson* where the *age*, *income*, *share*, *expenditure*, and *owner* are significant.

6.2.3 Models for Underdispersed Count Data

As presented for the *Smarket* data, we found an underdispersion since the variance is 0.1298 and the mean is 1.4783. The variance is only 0.0878 times the mean. This phenomenon is called underdispersion in counts regression and the *Quasi-Poisson regression* can be used to model such data. Note that the *negative-binomial regression* is not appropriate anymore since it is designed for overdispersed data.

As a comparison, we will fit both the conventional *Poisson regression* and the *quasi-Poisson regression* using *glm* as follows:

```
# quasi-Poisson regression
Smarket.quasipoisson = glm(Volume~ Lag1+Lag2+Lag3+Lag4+Lag5,
                           data=Smarket,
                           family = c("quasipoisson"))
# summary
summary(Smarket.quasipoisson)
```

```
##
## Call:
## glm(formula = Volume ~ Lag1 + Lag2 + Lag3 + Lag4 + Lag5,
## family = c("quasipoisson"), data = Smarket)
##
## Coefficients:
##             Estimate Std. Error t value Pr(>|t|)
## (Intercept)  0.39072    0.00687   56.85   <2e-16 ***
## Lag1         0.00828    0.00603    1.37    0.170
## Lag2        -0.00943    0.00604   -1.56    0.119
## Lag3        -0.00941    0.00603   -1.56    0.119
## Lag4        -0.01078    0.00604   -1.78    0.075 .
## Lag5        -0.00514    0.00600   -0.86    0.392
## ---
## Signif. codes:
## 0 '***' 0.001 '**' 0.01 '*' 0.05 '.' 0.1 ' ' 1
##
## (Dispersion parameter for quasipoisson family
## taken to be 0.08725)
```

```
##
##     Null deviance: 106.75  on 1249  degrees of freedom
## Residual deviance: 105.84  on 1244  degrees of freedom
## AIC: NA
##
## Number of Fisher Scoring iterations: 4
```

```
# the conventional Poisson regression
Smarket.poisson = glm(Volume~ Lag1+Lag2+Lag3+Lag4+Lag5,
                      Smarket, family = c("poisson"))
# summary
summary(Smarket.poisson)
```

```
##
## Call:
## glm(formula = Volume ~ Lag1 + Lag2 + Lag3 + Lag4 + Lag5,
## family = c("poisson"), data = Smarket)
##
## Coefficients:
##             Estimate Std. Error z value Pr(>|z|)
## (Intercept)  0.39072    0.02327   16.79   <2e-16 ***
## Lag1         0.00828    0.02041    0.41     0.68
## Lag2        -0.00943    0.02046   -0.46     0.64
## Lag3        -0.00941    0.02042   -0.46     0.64
## Lag4        -0.01078    0.02046   -0.53     0.60
## Lag5        -0.00514    0.02031   -0.25     0.80
##
## (Dispersion parameter for poisson family taken to be 1)
##
##     Null deviance: 106.75  on 1249  degrees of freedom
## Residual deviance: 105.84  on 1244  degrees of freedom
## AIC: Inf
##
## Number of Fisher Scoring iterations: 4
```

As seen from the output, the dispersion parameter is estimated at 0.08725266 as seen by *(Dispersion parameter for quasipoisson family taken to be 0.08725266)* in the quasi-Poisson regression *Smarket.quasipossion* model. Whereas the dispersion parameter is defaulted to 1 in the classical Poisson regression *Smarket.poisson* as seen by *(Dispersion parameter for poisson family taken to be 1)*. Due to the estimated underdispersed parameter of 0.08725266, the standard errors of all the parameters in *Smarket.poisson* are scaled and adjusted by 0.08725266, which resulted in smaller standard errors in the quasi-Poisson regression *Smarket.quasipoisson*. With this adjustment, the p-values are also reduced. However, there is still no statistical significance for all the covariates of *Lag1* to *Lag5* indicating that the percentage of the previous 5 days are not statistically significant predictors for the *Volume* (the volume of shares traded) and other predictors should be included in this prediction model. This makes sense since the stock markets are so complicated and there are so many competing predictors we have not included in this modeling.

6.3 Monte-Carlo Simulation Study

As discussed in previous section, *overdispersion/underdispersion*, a phenomenon frequently encountered in count data analysis, occurs when the observed variance of a dataset is larger/less than what would be expected based on the assumed statistical Poisson model. This departure from the expected variance can arise due to unaccounted-for sources of variability or the presence of extra-binomial variation. *Overdispersion/underdispersion* can have significant impacts on the accuracy and reliability of statistical analysis of the count data. *Dispersion* can distort parameter estimates, standard errors, and hypothesis tests, leading to misleading conclusions. Addressing *overdispersion/underdispersion* is imperative to ensure the validity of the results and the robustness of the analysis.

To explore the effects of *dispersion* on count data analysis, this section is aimed to incorporate its presence into a Monte-Carlo simulation study. By systematically introducing varying levels of *dispersion*, we aim to illustrate how different degrees of *dispersion* impact the performance of the conventional *Poisson regression* and how the *quasi-Poisson regression* to incorporate and adjust the *overdispersion/underdispersion* for correct count data analysis as outlined in the previous section. We leave the Monte-Carlo simulation study for *negative-binomial regression* to the interested readers as an exercise to further understand the logic behind this investigation.

Specifically, we will simulate *dispersed quasi-Poisson* count data in three scenarios with the first scenario for *no dispersion* (i.e., $\phi = 1$) to validate that both the *Poisson regression* and the *quasi-Poisson regression* will give the exact the same statistical inference, 2) *overdispersion* (i.e., $\phi = 5$) to mimic the *CreditCard* data and 3) *underdispersion* (i.e., $\phi = 0.5$) to mimic the *Smarket* data.

The following *4-step procedure* will be used for all three scenarios:

- *Step 1: True Values Specification.* To start, we make use of the *CreditCard* data to select the *age* and *income* variables to be used for this Monte-Carlo simulation. We also make use of the corresponding parameter estimates for this simulation, i.e., the intercept β_0 is *b0 = 1.5*, β_1 for *age* is *b4age = 0.006*; and β_2 for *income* is *b4income = 0.03*.

- *Step 2: Data Generation.* With the true parameters and the independent variables of *age* and *income* from *Step 1*, we can construct the linear predictor $\eta = \beta_0 + \beta_1 \times age + \beta_2 \times income$ = b0 + b4age×age + b4income×income, then we can get the *mean* for the *Poisson count* as $\mu = e^{\eta} = mean\_count$. The *dispersed* counts can then be generated by using *R* function *rpois* as $rpois(n, lambda = mean\_count) * dispersion$, where the initial *rpois(n, lambda = mean_count)* is used to generate the standard *Poisson* counts and then multiplying by *dispersion* to produce the *dispersed Poisson* counts. Depending on whether the *dispersion* parameter is a whole number of decimal number, the generated data could be non-integer number (if *dispersion* is a decimal value) which would not be appropriate for *Poisson regression* for *count* data. In this case, the generated data should be *rounded* to the integer counts before fitting the data to the *Poisson regression.*
- *Step 3: Fit Poisson and quasi-Poisson Regressions.* With the data generated from *Step 2*, we can then fit the *Poisson regression* and the *quasi-Poisson regression* to keep track the estimated parameters and standard errors to construct the associated 95% confidence intervals.
- *Step 4: Summarize the Simulation.* To conclude the procedure, *Step 1* through *Step 3* are repetitively executed for a substantial number of iterations, denoted as *nsimu* (set at, for instance, 1,000 iterations). In this step, we summarize the Monte-Carlo simulations with respect to the parameter estimation and coverage probability to compare the performance between the *Poisson regression* and *quasi-Poisson regression.*

The following sections will be used for a comprehensive explanation of the aforementioned steps with *R* code.

6.3.1 Scenerio 1: No Dispersion

In this scenario, the *dispersion* parameter $\phi = 1$. We will run the simulations for 1,000 times to keep tracking of 9 items:

- *Parameter Estimates* (3 items denoted by *b0, b4age, b4income*). There are six parameters with three estimated from the *Poisson regression* and three estimated from the *quasi-Poisson regression.* Due to the three parameters from both regressions being the same, we only need to keep track of three of them.
- *Coverage for Poisson Regression* (3 items denoted by *CP4b0PR,CP4bagePR, CP4bincomePR* for the three parameters). Using the estimated parameters and the associated standard errors from the *Poisson Regression,* we can construct the 95% confidence interval as $\hat{\beta} \pm 1.96 \times SE\left(\hat{\beta}\right)$ for $\beta = (\beta_0, \beta_1, \beta_2)$ = (b0, b4age, b4income). With these confidence intervals, we can then verify whether the true parameters $\beta = (\beta_0, \beta_1, \beta_2) = c(1.5, 0.006, 0.03)$ are covered within these confidence intervals.

- *Coverage for quasi-Poisson Regression*(3 items denoted by *CP4b0qPR*, *CP4bageqPR*,*CP4bincomeqPR* for the three parameters). The same explanation as above.

This is detailed in the following *R* implementation:

```
# Set random seed for reproducibility
set.seed(3388)
#
# Step 1: Take the Credit Card data as simulation setting
#
# Number of observations
n = dim(CreditCard)[1]
# Take the age and income from CreditCard
age = CreditCard$age; income = CreditCard$income
# Setting for True parameters
b0 = 1.5; b4age = 0.006; b4income= 0.03
TruePar = c(b0,b4age, b4income)
#
# Step 2: Simulate Data
#
# Get the linear predictor
eta = b0+b4age*age+b4income*income
# Mean count value
mean_count = exp(eta);
# Simulation now
nsimu = 1000
# Create a matrix to hold all the simulation results
estParMat = matrix(0, ncol=9, nrow=nsimu)
colnames(estParMat) = c("b0","b4age","b4income",
          "CP4b0PR","CP4bagePR","CP4bincomePR",
          "CP4b0qPR","CP4bageqPR","CP4bincomeqPR")
# Start the looping
for(s in 1:nsimu){
# Simulate the dispersed data
simu_counts = rpois(n, lambda = mean_count)
#
# Step 3: Model fitting
#
# Fit Poisson regression
mod.poisson = glm(simu_counts~age+income,
             family = c("poisson"))
# Extract the estimates
estPar    = summary(mod.poisson)$coef[,1]
SE.estPar = summary(mod.poisson)$coef[,2]
# Construct the coverage for Poisson regression
```

```
covered = (estPar-1.96*SE.estPar < TruePar)&(estPar+1.96*SE.estPar
> TruePar)
# Track the estimates and coverage
estParMat[s, 1:3] = estPar
estParMat[s, 4:6] = covered
# Fit the quasi-Poisson regression
mod.quasipoisson = glm(simu_counts~age+income,
             family = c("quasipoisson"))
# Extract the estimates
estPar     = summary(mod.quasipoisson)$coef[,1]
SE.estPar = summary(mod.quasipoisson)$coef[,2]
covered = (estPar-1.96*SE.estPar < TruePar)
& (estPar+1.96*SE.estPar > TruePar)
estParMat[s, 7:9] =  covered
} # end of s-looping
# Check the dimension: 1,000 by 9
dim(estParMat)
```

```
## [1] 1000    9
```

With the above 1,000 simulations, we can make use of *R* function *apply* to summarize (i.e., *average* them) the simulation results as follows:

```
#Step 4: Summarize the simulation
apply(estParMat,2,mean)
```

```
##        b0         b4age      b4income    CP4b0PR     CP4bagePR
## 1.497172       0.006039     0.030154    0.947000     0.946000
##  CP4bincomePR       CP4b0qPR     CP4bageqPR  CP4bincomeqPR
##     0.956000       0.947000       0.945000        0.954000
```

Observing the outcomes, it is evident that the computed *means* for the three parameters from these 1,000 simulations are at *(b0, b4age, b4income)* = (1.497172443, 0.006039049, 0.030153823), showcasing remarkable proximity to the true parameter values of (1.5, 0.006, 0.3). Furthermore, the calculated 95% coverage probabilities for these three parameters using the *Poisson regression* are *(CP4b0PR, CP4bagePR, CP4bincomePR)* = (0.947, 0.946, 0.956), while those derived from the *quasi-Poisson regression* as *(CP4b0qPR, CP4bageqPR, CP4bincomeqPR)* = (0.947, 0.945, 0.954), respectively. There are also close to the required 95%.

Should you possess a more powerful computing system, expanding the simulation to a greater number of iterations (e.g., 100,000) is anticipated to yield even better results.

This simulation can be graphically summarized in Figure 6.5 using *R* code chunk as follows:

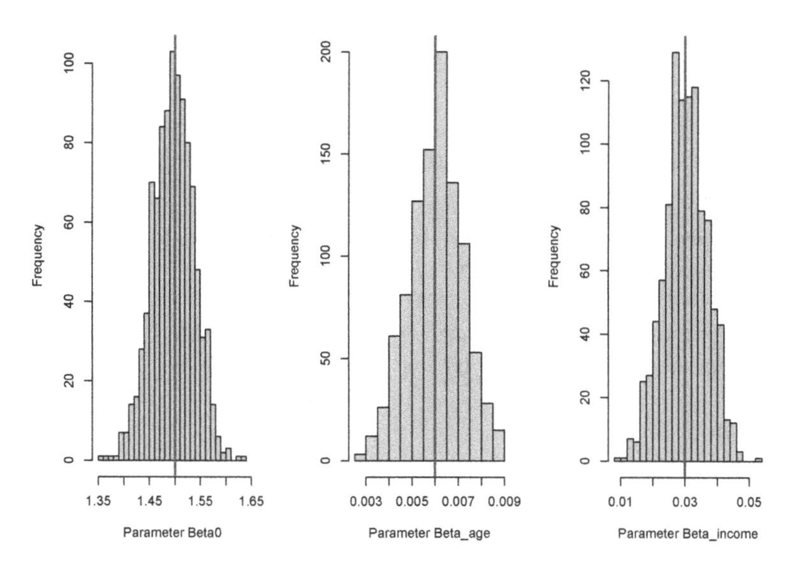

FIGURE 6.5
Monte-Carlo Distributions Overlaid with the True Parameters for Scenario 1

```
# Figure layout
par(mfrow=c(1,3))
# Make the plots for intercept
hist(estParMat[,1],nclass=20, main="",
     xlab="Parameter Beta0")
abline(v=b0, lwd=2, col="red")
hist(estParMat[,2], nclass=20,main="",
     xlab="Parameter Beta_age")
abline(v=b4age, lwd=2, col="red")
hist(estParMat[,3], nclass=20,main="",
     xlab="Parameter Beta_income")
abline(v=b4income, lwd=2, col="red")
```

6.3.2 Scenario 2: Overdispersion

We can now easily extend the above *Scenario 1* to include *overdispersion* in *Scenario 2* where the *dispersion* parameter $\phi = 5$. In this scenario, three changes are needed to make from *Scenario 1*:

1. *Data Generation.* The simulated counts should incorporate the *overdispersion.* This can be implemented in *R* code as $simu\_counts = rpois(n, lambda = mean\_count) * dispersion$.

2. *Estimation for ϕ* (2 additional items denoted by *estPhi*, *obsPhi* to be tracked). The *estPhi* is to denote the estimated *dispersion* from the *quasi-Poisson regression* and *obsPhi* is to denote the *dispersion* calculated from the simulated count data. These two items are used to compare with the simulated true *dispersion* parameter.

3. *Estimation of Intercept.* The estimated intercept from both regression are biased and should be theoretically adjusted. Since for *dispersed* count data ϕX, $E(\phi X) = \phi E(X) = \phi e^{\eta} = \phi e^{\beta_0+\beta_1 \times age+\beta_2 \times income} = e^{\beta_0+log(\phi)+\beta_1 \times age+\beta_2 \times income}$, therefore the estimated intercept parameter from the conventional *Poisson regression* should be adjusted by $\log(\phi)$ to make it correct.

The *R* implementation is as follows:

```
# Set random seed for reproducibility
set.seed(3388)
#
# Step 1: Take the Credit Card data as simulation setting
#
# Number of observations
n = dim(CreditCard)[1]
# Take the age and income from CreditCard
age = CreditCard$age; income = CreditCard$income
# Setting for True parameters
b0 = 1.5; b4age = 0.006; b4income= 0.03
TruePar = c(b0,b4age, b4income)
#
# Step 2: Simulate Data
#
# Get the linear predictor
eta = b0+b4age*age+b4income*income
# Mean count value
mean_count = exp(eta);
# Dispersion parameter for overdispersion
dispersion = 5
# Simulation now
nsimu = 1000
# Create a matrix to hold all the simulation results
estParMat = matrix(0, ncol=11, nrow=nsimu)
colnames(estParMat) = c("b0","b4age","b4income",
        "CP4b0PR","CP4bagePR","CP4bincomePR",
        "CP4b0qPR","CP4bageqPR","CP4bincomeqPR",
        "estPhi","obsPhi" )
# Start the looping
for(s in 1:nsimu){
```

```
# Simulate the dispersed data
simu_counts = rpois(n, lambda = mean_count )*dispersion
#
# Step 3: Model fitting
#
# Fit Poisson regression
mod.poisson = glm(simu_counts~age+income,
              family = c("poisson"))
# Extract the estimates and adjust the intercept
estPar    = summary(mod.poisson)$coef[,1] - c(log(dispersion),0,0)
SE.estPar = summary(mod.poisson)$coef[,2]
# Construct the coverage for Poisson regression
covered = (estPar-1.96*SE.estPar < TruePar)&(estPar+1.96*SE.estPar
 > TruePar)
# Track the estimates and coverage
estParMat[s, 1:3] = estPar
estParMat[s, 4:6] = covered
# Fit the quasi-Poisson regression
mod.quasipoisson = glm(simu_counts~age+income,
              family = c("quasipoisson"))
# Extract the estimates
estPar    = summary(mod.quasipoisson)$coef[,1]
SE.estPar = summary(mod.quasipoisson)$coef[,2]
covered = (estPar-1.96*SE.estPar < TruePar+c(log(dispersion),0,0))
 & (estPar+1.96*SE.estPar > TruePar+c(log(dispersion),0,0))
estParMat[s, 7:9] =  covered
# The dispersion parameters
estParMat[s,10] <- summary(mod.quasipoisson)$dispersion
# Calculate the dispersion parameter observed
data.phi = var(simu_counts)/mean(simu_counts);
estParMat[s,11] <- data.phi

} # end of s-looping
# Check the dimension: 1,000 by 11
dim(estParMat)
```

```
## [1] 1000   11
```

With the above 1,000 simulations, we can make use of *R* function *apply* to summarize the simulation results as follows:

```
#Step 4: Summarize the simulation
apply(estParMat,2,mean)
```

```
##            b0          b4age        b4income    CP4b0PR
##      1.497172       0.006039        0.030154   0.616000
##     CP4bagePR   CP4bincomePR     CP4b0qPR    CP4bageqPR
##     0.611000       0.619000      0.947000     0.945000
##     CP4bincomeqPR     estPhi          obsPhi
##       0.954000      4.995106        5.274204
```

It can be seen from the above summary, the computed *means* for the three parameters from these 1,000 simulations are at *(b0, b4age, b4income)* = (1.497172443, 0.006039049,0.030153823). Notice that the estimated intercept from the conventional *Poisson regression* is adjusted by the *log(5) = 1.609438*. With this adjustment, all three parameters are again showing remarkable proximity to the true parameter values of (1.5, 0.006, 0.3). Furthermore, the estimated *dispersion* $\phi = 4.995105595$ which is close to the true ϕ of 5 along with the calculated ϕ from the simulated data as 5.274203726.

The resulted 95% coverage probabilities for these three parameters using the *Poisson regression* are *(CP4b0PR, CP4bagePR, CP4bincomePR)* = (0.616, 0.611, 0.619), while those obtained from the *quasi-Poisson regression* as *(CP4b0qPR, CP4bageqPR, CP4bincomeqPR)* = (0.947, 0.945, 0.954), respectively. Evidently, the *quasi-Poisson regression* consistently maintains the correct 95% coverage probability. However, it is notable that the conventional *Poisson regression* yields coverage probabilities lower than the intended 95%, indicative of erroneously *underestimated* standard errors. This scenario could potentially lead to *Type-I errors* and an inflation of reported significance with higher level of *false positive.*

This simulation can be graphically summarized in Figure 6.6 using the *R* code chunk as follows where we can see that the means for the estimated three regression parameters along with the *dispersion* parameters from the *quasi-Poisson regression* are close to the true values:

```
par(mfrow=c(2,3))
# Make the plots for logistic regression model
hist(estParMat[,1],nclass=20, main="",
     xlab="Parameter Beta0")
abline(v=b0, lwd=2, col="red")
hist(estParMat[,2], nclass=20,main="",
     xlab="Parameter Beta_age")
abline(v=b4age, lwd=2, col="red")
hist(estParMat[,3], nclass=20,main="",
     xlab="Parameter Beta_income")
abline(v=b4income, lwd=2, col="red")
# Make the plots for linear regression model
hist(estParMat[,10], nclass=20, main="",
     xlab="Estimated Dispersion:Phi")
```

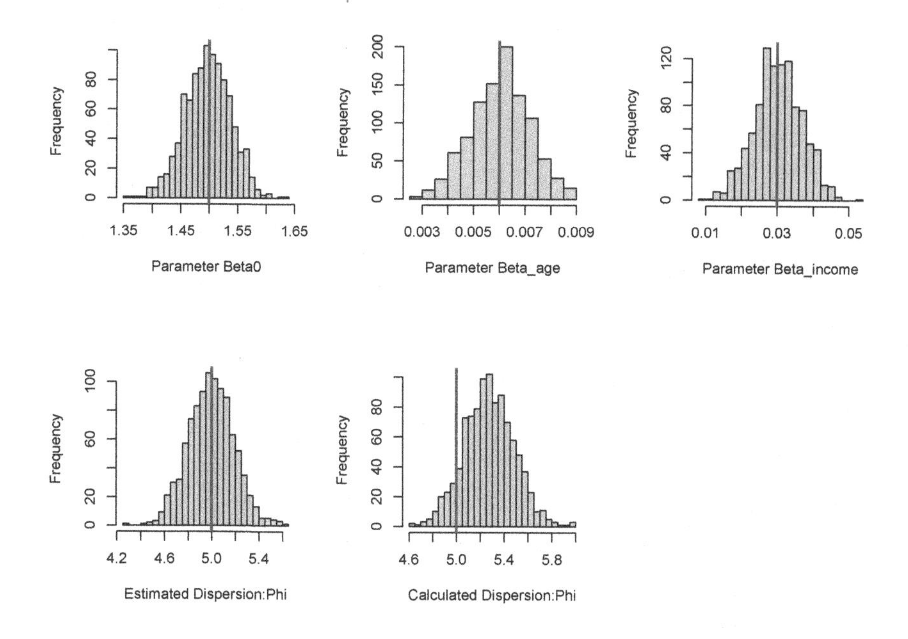

FIGURE 6.6
Monte-Carlo Distributions Overlaid with the True Parameters for Scenario 2

```
abline(v=dispersion, lwd=2, col="red")
hist(estParMat[,11], nclass=20,  main="",
    xlab="Calculated Dispersion:Phi")
abline(v=dispersion, lwd=2, col="red")
```

6.3.3 Scenario 3: Underdispersion

To consider *underdispersion* in *Poisson regression*, we can modify the above *Scenario 2* to include *underdispersion* for *Scenario 3* where the *dispersion* parameter $\phi = 0.5$. This *dispersion* parameter is not integer, but a decimal number of 0.5. The simulated counts can then be non-integers. To incorporate this *underdispersion* with decimal values, we can use *R* function *round* to round them to integers using the follow *R* code: $simu\_counts = round(rpois(n, lambda = mean\_count) * dispersion)$.

The rest of *R* implementation is exactly the same as what we did in *Scenario 2*. This is showcased as below:

```
# Set random seed for reproducibility
set.seed(3388)
```

```
#
# Step 1: Take the Credit Card data as simulation setting
#
# Number of observations
n = dim(CreditCard)[1]
# Take the age and income from CreditCard
age = CreditCard$age; income = CreditCard$income
# Setting for True parameters
b0 = 1.5; b4age = 0.006; b4income= 0.03
TruePar = c(b0,b4age, b4income)
#
# Step 2: Simulate Data
#
# Get the linear predictor
eta = b0+b4age*age+b4income*income
# Mean count value
mean_count = exp(eta);
# Dispersion parameter underdispersion
dispersion = 0.5
# Simulation now
nsimu = 1000
# Create a matrix to hold all the simulation results
estParMat = matrix(0, ncol=11, nrow=nsimu)
colnames(estParMat) = c("b0","b4age","b4income",
            "CP4b0PR","CP4bagePR","CP4bincomePR",
            "CP4b0qPR","CP4bageqPR","CP4bincomeqPR",
            "estPhi","obsPhi" )
# Start the looping
for(s in 1:nsimu){
# Simulate the dispersed data
simu_counts = round(rpois(n, lambda = mean_count )*dispersion)
#
# Step 3: Model fitting
#
# Fit Poisson regression
mod.poisson = glm(simu_counts~age+income,
            family = c("poisson"))
# Extract the estimates
estPar    = summary(mod.poisson)$coef[,1]-c(log(dispersion),0,0)
SE.estPar = summary(mod.poisson)$coef[,2]
# Construct the coverage for Poisson regression
covered = (estPar-1.96*SE.estPar < TruePar)
 & (estPar+1.96*SE.estPar > TruePar)
# Track the estimates and coverage
```

```
estParMat[s, 1:3] = estPar
estParMat[s, 4:6] = covered
# Fit the quasi-Poisson regression
mod.quasipoisson = glm(simu_counts~age+income,
            family = c("quasipoisson"))
# Extract the estimates
estPar    = summary(mod.quasipoisson)$coef[,1]
SE.estPar = summary(mod.quasipoisson)$coef[,2]
covered = (estPar-1.96*SE.estPar < TruePar+c(log(dispersion),0,0))
 &  (estPar+1.96*SE.estPar > TruePar+c(log(dispersion),0,0))
estParMat[s, 7:9] =  covered
# The dispersion parameters
estParMat[s,10] <- summary(mod.quasipoisson)$dispersion
# Calculate the dispersion parameter observed
data.phi = var(simu_counts)/mean(simu_counts);
estParMat[s,11] <- data.phi

} # end of s-looping
# Check the dimension: 1,000 by 11
dim(estParMat)

## [1] 1000   11
```

With the above 1,000 simulations, we can make use of *R* function *apply* to summarize the simulation results as follows:

```
#Step 4: Summarize the simulation
apply(estParMat,2,mean)

##          b0       b4age     b4income
##    1.498261    0.006013     0.030119
##      CP4b0PR   CP4bagePR  CP4bincomePR
##    0.993000    0.998000     0.997000
##     CP4b0qPR  CP4bageqPR CP4bincomeqPR
##    0.947000    0.944000     0.955000
##      estPhi      obsPhi
##    0.537668    0.565234
```

Again, the computed *means* for the three parameters from these 1,000 simulations are at *(b0, b4age, b4income)* = (1.498260543, 0.006012589, 0.030119157). Notice that again the estimated intercept from the conventional *Poisson regression* is adjusted by the *log(0.5)* = *-0.6931472*. With this adjustment, all three parameters are again showing remarkable proximity to the true parameter values of (1.5, 0.006, 0.3). Furthermore, the estimated *dispersion* ϕ = 0.537668036 which is close to the true ϕ of 0.5 along with the calculated ϕ from the simulated data as 0.565233909.

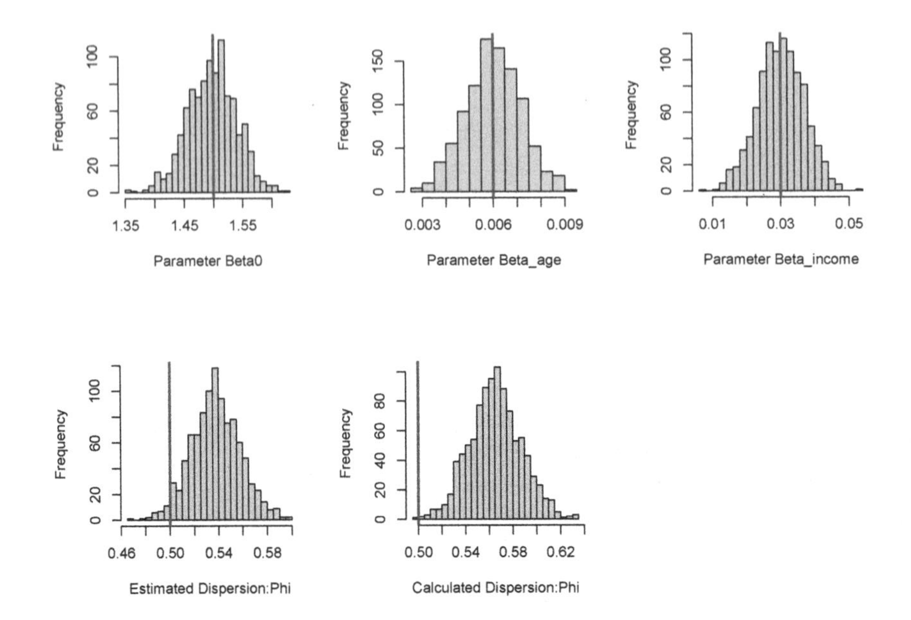

FIGURE 6.7
Monte-Carlo Distributions Overlaid with the True Parameters for Scenario 3

The resulted 95% coverage probabilities for these three parameters using the *Poisson regression* are *(CP4b0PR, CP4bagePR, CP4bincomePR)* = (0.993,0.998, 0.997), while those obtained from the *quasi-Poisson regression* as *(CP4b0qPR, CP4bageqPR, CP4bincomeqPR)* = (0.947, 0.944, 0.955), respectively. Evidently, the *quasi-Poisson regression* consistently maintains the correct 95% coverage probability. However, it is notable that the conventional *Poisson regression* yields coverage probabilities higher than the intended 95%, indicative of erroneously *overestimated* standard errors. This scenario could potentially lead to *Type-II errors* and a deflation of reported significance with higher-level of *false negative.*

This simulation can be graphically summarized in Figure 6.7 using the *R* code chunk as follows where we can see that the means for the estimated three regression parameters along with the *dispersion* parameters from the *quasi-Poisson regression* are close to the true values:

```
par(mfrow=c(2,3))
# Make the plots for logistic regression model
hist(estParMat[,1],nclass=20, main="",
     xlab="Parameter Beta0")
abline(v=b0, lwd=2, col="red")
```

```
hist(estParMat[,2], nclass=20,main="",
     xlab="Parameter Beta_age")
abline(v=b4age, lwd=2, col="red")
hist(estParMat[,3], nclass=20,main="",
     xlab="Parameter Beta_income")
abline(v=b4income, lwd=2, col="red")
# Make the plots for linear regression model
hist(estParMat[,10], nclass=20, main="",
     xlab="Estimated Dispersion:Phi")
abline(v=dispersion, lwd=2, col="red")
hist(estParMat[,11], nclass=20,  main="",
     xlab="Calculated Dispersion:Phi")
abline(v=dispersion, lwd=2, col="red")
```

6.4 Discussions

This chapter introduced the *Poisson regression* to model count data in financial data analysis. We used two datasets with the first dataset *CreditCard* to illustrate the conventional *Poisson regression* with *quasi-Poisson regression* and *negative-binomial regression*, since this data is *overdispersed*. We used the second dataset *Smarket* from Chapter 5 to compare the conventional *Poisson regression* and the *quasi-Poisson regression* to model the outcome variable *Volume* (the volume of shares traded in billions) since this outcome variable *underdispersed* data.

To investigate the effects of *overdispersion* and *underdispersion* on the conventional *Poisson regression*, we additionally designed and conducted a Monte-Carlo simulation study. From this investigation, we found that *overdispersion* can lead to Type-I errors by producing an excessive number of false-positive significances in the conventional *Poisson regression*, and *underdispersion* can lead to Type-II errors by producing an excessive number of false-negative significances. However, the *quasi-Poisson regression* can mitigate the effects from *overdispersion* and *underdispersion* to produce correct statistical parameter estimations, confidence intervals, and the associated statistical inference.

As a recommendation in analyzing count data in financial analysis, do make sure that the data meets the assumptions of a *Poisson regression*. If there exists an *overdispersion*, make use of the *quasi-Poisson regression* and/or *negative-binomial regression*. And if there exists *underdispersion*, make use of the *quasi-Poisson regression*.

Poisson regression can provide insights into the relationships between predictor variables and count-based financial phenomena. However, keep in mind that financial data can be complex and influenced by various factors, so model selection and interpretation should be done with careful consideration of the domain context. It is always important to carefully interpret results and consider the economic or financial significance of the coefficients.

Remember that while *Poisson regression* can be a useful tool for financial modeling, it's essential to consider the limitations and assumptions of the model, as well as the complexities of financial market data. Additionally, you might want to explore other modeling approaches, such as time series models (e.g., ARIMA) to capture more intricate relationships within the data, which are the topics for the next chapters.

6.5 Exercises

Since the three variables of *selfemp*, *dependents* and *months* are not statistically significant in both *Poisson Regression* and *quasi-Poisson Regression*, remove these three variables and refit the *Poisson Regression* and *quasi-Poisson Regression*. Answer the following questions:

1. Do you find better model fitting?
2. Are the variables *age*, *income*, *share*, *expenditure*, and *owner* more significant in the new regressions?
3. Is there still *overdispersion* in the *Poisson regression*?

7

Autoregressive Integrated Moving-Average Models

As demonstrated in Chapter 2, we briefly explored the application of *linear regression* models to analyze financial time series data. However, *linear regression* models assume that data or residuals are *independent* of each other. This assumption means that each observation is not influenced by its preceding or succeeding observations. However, this assumption often breaks down when dealing with time series data, as financial data frequently exhibit time-dependent patterns known as *autocorrelation.*

Autocorrelation refers to the correlation of a variable with its own past values at different lags. In financial time series data, it is common to observe *autocorrelation*, as asset prices and market indicators often depend on their historical values. This *time-dependent* structure can lead to violations of the independence assumption underlying the *linear regression.*

This chapter then serves as an introduction to *ARIMA(p,d,q) Regression* as a solution to the challenges posed by time series data. ARIMA stands for *Autoregressive Integrated Moving-Average.* It is characterized by three main components: 1) *Autoregressive (AR)* component represented by *order* p to capture the relationship between the financial time series data and its past values, 2) *Integrated* component represented by *order* d to denote the differencing degree needed to make the data *stationary* (i.e., remove trends), and 3) *Moving-Average (MA)* component represented by *order* q to model the dependency on past forecast errors.

Throughout this introduction, we continually utilize the *wages* data used in Chapter 2. This dataset effectively serves as a foundation for the discussions and demonstrations in this chapter as well as a real-world example to illustrate the application of *ARIMA regression* techniques in the analysis of time series data.

To conclude the chapter, we design and implement a Monte-Carlo simulation study. This study aims to highlight the differences in conclusions that can arise when analyzing financial time series data using *linear regression* versus *ARIMA regression.* It underscores the importance of choosing the appropriate modeling technique when dealing with time-dependent data.

DOI: 10.1201/9781003469704-7

Overall, this chapter focuses on the transition from *linear regression* to *ARIMA regression* in the context of financial time series analysis, addressing the challenges posed by *autocorrelation* and demonstrating the practical application of *ARIMA* techniques using real data and simulation studies.

7.1 Autocorrelation Function and Partial Autocorrelation Function

The first and the most foundational concept in understanding time series analysis is the *autocorrelation* and *partial autocorrelation* function.

7.1.1 Autocorrelation Function

Time series *autocorrelation* is the correlation between two observations at different time points in a time series. It is the correlation for observations with observations from previous time steps (called *lags*). For example in the *wages* dataset, the wages at year t are correlated with the wages at previous (*lagged*) years $t-k$ where $k = 1, 2, 3, \ldots$. The closer the time interval, the higher the correlation between the *wages*. When these correlations are present, the values in the past would influence the value at current.

Therefore, *autocorrelation* is to measure the correlation between two values in a time series. Because the correlation of the time series observations is calculated with values of the same series at previous times, this is called a *serial correlation*, or an *autocorrelation*. In other words, the time series data correlate with themselves and hence, the name of *autocorrelation*. In statistical terms, the observations at y_t and y_{t-k} are separated by k time units, where k is called the *lag*. When $k = 1$, we assess adjacent observations. For each lag, there is a correlation.

The *autocorrelation function (ACF)* is then designed to assess the correlation between observations in a time series for a set of lags. This *ACF* for time series y_t $(t = 1, 2, \ldots)$ is defined as

$$\gamma(k) = cor(y_t, y_{t-k}) \tag{7.1}$$

for k = 1,2,.... The *ACF*s can be found and plotted using *R* function *acf*, which we will illustrate in the analysis of *wage* data.

7.1.2 Partial Autocorrelation Function

In addition to the *autocorrelation function*, the *partial autocorrelation function (PACF)* summarizes the correlation between observations and their prior time steps, with the effects of intervening observations removed. It displays only the correlation between two observations that the shorter lags between those observations do not explain. For example, the *partial autocorrelation* for lag 3 is only the correlation that lags 1 and 2 do not explain. That is to say, the partial autocorrelation at lag k is the correlation that results after removing the effect of any correlations due to the terms at shorter lags. More specifically, given a time series y_t, the partial autocorrelation of lag k (denoted $\phi(k)$) is the autocorrelation between y_t and y_{t+k} with the linear dependence of y_t on y_{t+1} to y_{t+k-1} removed. Equivalently, this is the *autocorrelation* between y_t and y_{t+k} that is not accounted for by lags 1 through $k-1$. The PACF can be useful for identifying the order of an AR process.

The difference between ACF and $PACF$ can be easily illustrated from the regression model perspective. To estimate the ACF $\gamma(k)$ at lag k, we can fit a simple linear regression model to the time series y_t as follows:

$$y_t = \alpha_0 + \alpha_k y_{t-k} + u_t \tag{7.2}$$

With this linear regression model, the estimated slope parameter is the estimated ACF, i.e., $\hat{\alpha}_k = \gamma(k)$. Based on the regression theory and if you expand the parameter estimation in equation (2.7). it is easy to see that the estimate α_k is the sample correlation between y_t and y_{t-k}. as defined in equation (7.1).

To estimate the $PACF$ $\phi(k)$ at lag k, we can fit a multiple linear regression model as follows:

$$y_t = \alpha_0 + \alpha_1 y_{t-1} + \cdots + \alpha_{t-k+1} y_{t-k+1} + \alpha_k y_{t-k} + u_t \tag{7.3}$$

and the resulting $\hat{\alpha}_k$ is the estimate of $PACF$ of ϕ_k. Since this estimate is estimated with controlling all the intermediate elements from y_{t-1} to y_{t-k+1}, therefore, the $\phi(k)$ is the correlation between y_t and y_{t-h} after controlling for the intermediate elements.

The $PACF$s can be calculated and plotted using R function *pacf*, which we will illustrate in the analysis of *wage* data.

7.1.3 Ljung-Box Test for Significance

The $ACF/PACF$s are typically displayed in a plot with lags. In order to show the statistical significance, this $ACF/PACF$ plot is usually plotted with test bounds. These bounds are used to test the null hypothesis that an *autocorrelation coefficient* is 0. The null hypothesis is rejected if the sample autocorrelation is outside the bounds.

The *Ljung–Box* test is the typical test used in this case. The null hypothesis of this test is:

$$H_0 : \gamma(1) = \gamma(2) = \cdots = \gamma(p) = 0$$

for some lag-p. If the *Ljung–Box* test rejects the null, then we conclude that one or more of $\gamma(1), \gamma(2), \ldots, \gamma(p)$ is nonzero. In summary, the Ljung-Box test is test for the *absence* of autocorrelation. A low *p-value* below a given significance level indicates the values are autocorrelated. If the *p-value* > *0.05*, this implies that the time series data are are independent.

The bounds for *Ljung–Box test* are typically plotted with the *ACF/PACF* in *R* functions *tsdiag*, which will plot the residuals, the autocorrelation function of the data and the p-values of the Ljung-Box Test for all lags. This is illustrated in the analysis of *wage* dataset below.

7.1.4 Illustration with *Wages* Data

To graphically depict the *ACF* and *PACF* for a time series, we plot the *ACF/PACF*s with each bar representing the size and direction of the correlation. *Bars that extend across the red line are statistically significant.*

In the case that the time series are not correlated, *ACF/PACF* should be near zero for all lags, which we usually refer to as *white noise*. If there is at least one significant lag, it would indicate that a time series analysis, instead of the classical linear regression, should be used to incorporate lags into a regression analysis to model the data appropriately.

To see if autocorrelation exists in the *wages* data, we can call the *R* function *acf* and *pacf* to plot the autocorrelation function and partial autocorrelation for this data. This is shown in the following *R* code chunk:

```
# Extract the wage data
w = nwages[,"w"]
# Figure layout
par(mfrow=c(2,1),mar=c(4,4,1,2),oma=c(1,1,1,1) )
# Call acf function to plot the autocorrlation
acf(w,xlab="Time Series Lags", ylab="ACF", main="")
# Call pacf function to plot the partial autocorrlation
pacf(w,xlab="Time Series Lags", ylab="PACF", main="")
```

As seen from Figure 7.1, there is a significant time series autocorrelation to lag 3 from the ACF plot (i.e., out of Ljung-Box bounds as illustrated in the blue dashed horizontal lines). Also there is a significant lag-1 from *PACF*s. This is an indication that the *wage* data is a time series data with significant *autoregressive (AR) process*, which we will discuss later.

Given the pronounced *autocorrelation* present within the dataset, the conventional *linear regression* model discussed in Chapter 2 could potentially exhibit bias due to the inherent time series dependence. Consequently, to effectively

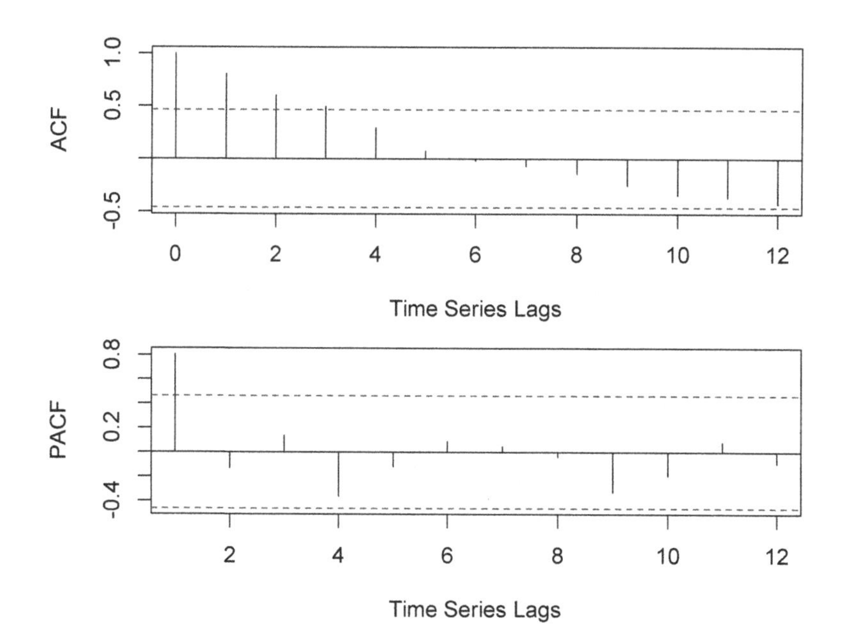

FIGURE 7.1
ACF and PACF in Wage Data

address this issue and account for the *autocorrelation*, a structured time series analysis is needed for the comprehensive analysis of this dataset.

7.2 Auto-Regressive Integrated Moving-Average Time Series

The foundational framework for time series analysis often begins with the modeling of *Auto-Regressive Integrated Moving-Average (ARIMA)* time series or processes. In the upcoming sections, we will introduce and provide a more detailed explanation of these models.

7.2.1 Auto-Regressive Process with Order p

The first model to model the time series autocorrelation as observed in Figure 7.1 is the auto-regressive model with order p, denoted by AR(p). In AR(p) process, we model y_t to its previous p values of the process, $y_{t-1}, \cdots, y_{t-p}$. More formally, a stochastic process y_t is an AR(p) process if:

$$y_t - \mu = \gamma_1(y_{t-1} - \mu) + \gamma_2(y_{t-2} - \mu) + \cdots + \gamma_p(y_{t-p} - \mu) + \sigma\epsilon_t, \qquad (7.4)$$

where μ is mean of the y_t, and the ϵ_t, $t = 1, 2, \cdots$ is white-noise (WN) process with mean 0 and variance σ_ϵ^2.

This AR(p) model is in fact a multiple linear regression model with the p-lagged values of the time series as the *x-variables*, which can be expressed as

$$y_t = \gamma_0 + \gamma_1 y_{t-1} + \gamma_2 y_{t-2} + \cdots + \gamma_p y_{t-p} + \epsilon_t \qquad (7.5)$$

where $\gamma_0 = \{1 - (\gamma_1 + \cdots + \gamma_p)\}\,\mu$. It can be shown that $1-(\gamma_1 + \cdots + \gamma_p) > 0$ for any stationary process, which will be discussed more in Section 7.5.

The parameters of $\left(\gamma_1, \cdots, \gamma_p, \mu, \sigma_\epsilon^2\right)$ associated with *AR(p)* can be estimated by the methods of *conditional least squares* or *maximum likelihood estimators*, which are programmed in most of the software, such as *R* with function *arima()*. The residuals are defined by

$$\hat{\epsilon}_t = y_t - \{\hat{\gamma}_0 + \hat{\gamma}_1 y_{t-1} + \hat{\gamma}_2 y_{t-2} + \cdots + \hat{\gamma}_p y_{t-p}\} \qquad (7.6)$$

for any $t > p + 1$.

These residuals are then used for model diagnostics with *residual autocorrelation* examined by the sample *ACF* along with the *Ljung–Box test*. For example, if an *AR(p)* model fits the time series data well, then the residuals should be like white noise. Any significant *residual autocorrelation* is a sign that the *AR(p)* model does not fit well.

7.2.2 Moving-Average Process with Order q

Another model used in time series is a moving-average (MA) model. A process y_t is a moving-average process with order of q, i.e., $MA(q)$, if y_t can be expressed as a weighted average (i.e., moving-average) of the past q values of the white noise process $\{\epsilon_t\}$.

More formally, the *MA(q)* process is defined as

$$y_t = \mu + \epsilon_t + \phi_1\epsilon_{t-1} + \cdots + \phi_q\epsilon_{t-q} \qquad (7.7)$$

Note that *MA(q)* is different from *AR(p)*, where *AR(p)* is to model y_t to the previous p-lagged values from y_{t-1} to y_{t-p}, where the *MA(q)* is to model y_t with the q-lagged error terms from ϵ_t to ϵ_{t-q}.

It is easy to show that $\gamma(k) = 0$ and $\phi(k) = 0$ if $|k| > q$. We can also use *R* function *arima* to fit MA(q) models.

7.2.3 ARMA Process with Order of p and q

Combine the *Auto-Regressive(AR)* process of order p (i.e., *AR(p)*) and the *Moving-Average (MA)* process of order q (i.e., *MA(q)*), we have an *ARMA(p,q)* process. This hybrid model can be used to capture more complex *autocorrelation* behavior in time series data when either a pure *AR(p)* or a pure *MA(q)* process alone cannot provide a good fit to the data.

The *ARMA(p,q)* model combines the *autoregressive* behavior (AR) that depends on past values of the series and the *moving-average* behavior (MA) that depends on past white noise or error terms. By incorporating both components, the *ARMA* models can often capture a broader range of *autocorrelation* patterns seen in real-world financial data.

To formally define an *ARMA(p, q)* model, we can use the notations in *AR(p)* as defined in equation (7.4) and the *MA(q)* as defined in (7.7) to define the *ARMA(p,q)* as follows:

$$\begin{aligned}(y_t - \mu) &= \gamma_1(y_{t-1} - \mu) + \cdots + \gamma_p(y_{t-p} - \mu) \\ &\quad + \epsilon_t + \phi_1\epsilon_{t-1} + \cdots + \phi_q\epsilon_{t-q}. \end{aligned} \tag{7.8}$$

As seen in equation (7.8), the y_t in *ARMA(p,q)* depends on the lagged values of itself to pth-order and lagged values of the white noise process to the qth-order. The parameter estimation of this *ARMA(p,q)* process can be done using *R* function *arima*.

7.2.4 Auto-Regressive Integrated Moving-Average Process

By adjusting the values of p and q, we can use the *ARMA* model to fit the specific autocorrelation patterns observed in the financial time series data. Further to the *ARMA(p,q)* model when the data exhibits *non-stationary* behavior (e.g., *trends or seasonality*), we may need to employ *differencing* to make it *stationary* before applying the *ARMA* model. This leads to more comprehensive *ARIMA(p,d,q) model*, which is the *auto-regressive integrated moving-average process* with d representing the *differencing order.*

To define the *ARIMA* process, let us define the *backward operator* (or *lag operator*) commonly used in time series analysis. A *backward operator* B is defined as

$$By_t = y_{t-1}$$

We can recursively apply this *backward operator* to a time series y_t multiple times, such as

$$B^d y_t = B^{d-1}(By_t) = B^{d-1}y_{t-1} = \cdots = y_{t-d}$$

Therefore, the *backward operator* B is to backup the time series by one unit and B^d will backup repeatedly for d times.

With the *backward operator*, the *ARMA(p,q)* process in equation (7.8) can be simplified and formulated as follows:

$$(1 - \gamma_1 B - \cdots - \gamma_p B^p)(y_t - \mu) = (1 + \phi_1 B + \cdots + \phi_q B^q)\,\epsilon_t. \tag{7.9}$$

Based on the *backward operator*, another useful *operator* in time series analysis is the *difference operator*, which is defined as

$$\Delta = 1 - B.$$

That is $\Delta y_t = y_t - By_t = y_t - y_{t-1}$. Similarly, the dth order difference $\Delta^d y_t = (1 - B)^d y_t = \sum_{i=0}^{d} \binom{d}{i}(-1)^i y_{t-i}$.

With the introduction of *difference operator* Δ, we can now define the *ARIMA* process. A time series y_t is an *ARIMA(p, d, q)* process if $\Delta^d y_t$ is an *ARMA(p, q)*. Notice that an *ARIMA(p, 0, q)* model is the same as an *ARMA(p, q)* model. Also an *ARIMA(p, 0, 0)*, *ARMA(p, 0)*, and *AR(p)* models are the same *auto-regressive* model with order p. Similarly, an *ARIMA(0, 0, q)*, *ARMA(0, q)*, and *MA(q)* models are the same *moving-average* with order q.

7.2.5 ARIMA with Regression

Recall the *multiple linear regression* model discussed in Chapter 2, the errors ϵ_t are assumed to be mutually independent. However, if the data $\{(X_t, y_t), t = 1, \cdots, n\}$ are time series with y_t as response variable and X_t as the matrix of K independent variables, it is highly likely that the errors are correlated. This correlation is a problem in regression analysis and can cause incorrect estimation of the standard errors and confidence intervals. Particularly, the coverage probability of confidence intervals can be much lower than the nominal value. A solution to this problem is to model the residuals as an ARIMA process.

Therefore, for a dataset with $\{(X_t, y_t), t = 1, \cdots, n\}$, we can define a linear regression model with *ARIMA* errors as follows:

$$y_t = \beta_0 + \beta_1 X_{t,1} + \cdots + \beta_K X_{t,K} + \sigma_\epsilon \epsilon_t, \tag{7.10}$$

where ϵ_t is an *ARIMA(p,d,q)* process; K is the number of independent variables in the data matrix X_t, σ_ϵ represents the standard deviation of y_t.

The regression parameters $(\beta_0, \beta_1, \cdots, \beta_K)$ along with the parameters associated with the *ARIMA(p,d,q)* can be estimated by the maximum likelihood estimation. This estimation is implemented in R function *arima* and we will illustrate the application of these models with real data.

7.3 Fitting *ARIMA* using *R*

To illustrate the time series analysis, we reuse the *wage* data. As seen from Figure 7.1, there is a significant time series autocorrelation to lag 3 from the *ACF* plot (i.e., out of Ljung-Box). This time series autocorrelation caused the dependence in the residuals in this regression as illustrated in Chapter 2.

To illustrate the regression using *ARIMA* model, we will make use of the *R* function *arima* to re-analyze this *wage* data.

Let us extract the data from *wage* dataset to be used for time series analysis as follows:

```
# Extract the data from nwage dataset
w    = nwages[,"w"];     CPI  = nwages[,"CPI"]
CPI1 = nwages[,"CPI1"]; CPI2 = nwages[,"CPI2"]
u   = nwages[,"u"];      mw   = nwages[,"mw"]
```

7.3.1 Fitting AR(p) Model

To fit *AR(p)* model, we need to select an optimal order of lags. In this case, we fit a series *AR(p)* model with $p = 1$, 2, 3, and 4 and select the best model for the data. This can be easily implemented in *R* as follows:

```
# Fit AR(p) with p=1 and print the model fitting
ar1 = arima(w, order = c(1,0,0));ar1
```

```
##
## Call: arima(x = w, order = c(1, 0, 0))
##
## Coefficients:
##         ar1  intercept
##       0.879      5.218
## s.e.  0.102      1.417
##
## sigma^2 estimated as 0.948: loglikelihood=-25.8, aic=57.61
```

```
# Fit AR(p) with p=2 and print the model fitting
ar2 = arima(w, order = c(2,0,0));ar2
```

```
##
## Call: arima(x = w, order = c(2, 0, 0))
```

```
##
## Coefficients:
##          ar1     ar2  intercept
##        0.932  -0.063      5.231
## s.e.   0.228   0.242      1.355
##
## sigma^2 estimated as 0.945: loglikelihood=-25.77, aic=59.54
```

```
# Fit AR(p) with p=3 and print the model fitting
ar3 = arima(w, order = c(3,0,0));ar3
```

```
##
## Call: arima(x = w, order = c(3, 0, 0))
##
## Coefficients:
##          ar1     ar2    ar3  intercept
##        0.946  -0.272  0.235      5.142
## s.e.   0.223   0.310  0.226      1.586
##
## sigma^2 estimated as 0.881: loglikelihood=-25.26, aic=60.52
```

```
# Fit AR(p) with p=4 and print the model fitting
ar4 = arima(w, order = c(4,0,0));ar4
```

```
##
## Call: arima(x = w, order = c(4, 0, 0))
##
## Coefficients:
##          ar1     ar2    ar3     ar4  intercept
##        1.039  -0.398  0.527  -0.326      5.275
## s.e.   0.217   0.312  0.297   0.236      1.145
##
## sigma^2 estimated as 0.787: loglikelihood=-24.37, aic=60.74
```

As seen from the model fitting, each model estimated the *AR(p)* model coefficients along with the standard error (i.e., s.e.), named as *ar1* for AR(1) model, *ar2* for *AR(2)* model, *ar3* for *AR(3)* model, and *ar4* for *AR(4)* model. The *intercept* in each model is the estimate of μ. In addition, we can see that there are output for model diagnostics as seen in $\hat{\sigma}_\epsilon^2$, along with the *log-likelihood* and the value of *aic*.

We can use the value of *aic* for model selection to see which model has the lowest AIC value. This can be done as follows:

```
# Output the AICs
AIC(ar1, ar2,ar3, ar4)
```

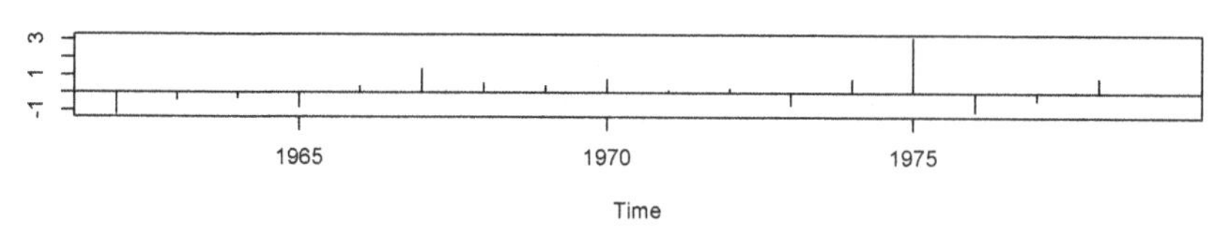

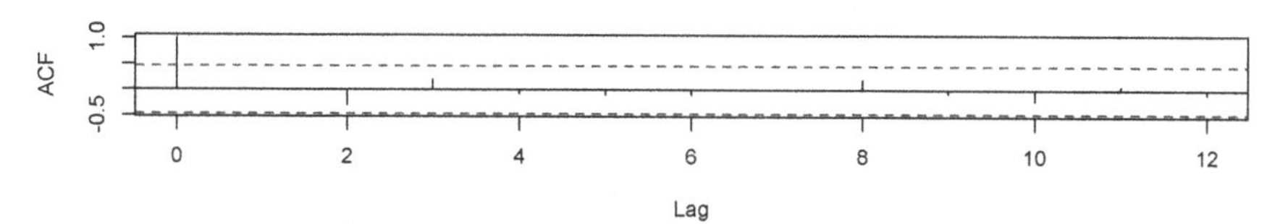

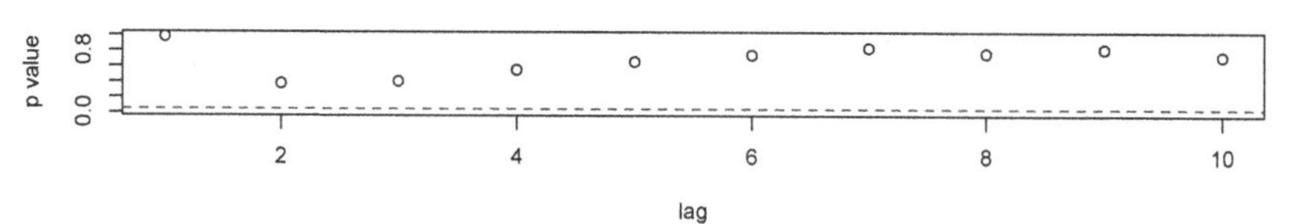

FIGURE 7.2
Model Diagnostics Plot for AR(1) Model

```
##      df   AIC
## ar1   3 57.61
## ar2   4 59.54
## ar3   5 60.52
## ar4   6 60.74
```

Among these four AR(p) models, the *AR(1)* model is the best with the lowest AIC = 57.605. That means the *wage* can be best modeled as an *auto-regressive order-1 process.*

For model diagnostics, we can make use of the *R* function *tsdiag* to examine the residuals after the *AR(1)* model fitting using the following *R* code chunk which will produce Figure 7.2 to graphically illustrate the plots for residual time series, ACF and the p-values for Ljung-Box statistic.

```
# Time series diagnostics
tsdiag(ar1)
```

As seen from the Figure 7.2, the *ACF*s are all within the bounds of Ljung-Box and the p-values are all large. Therefore, the *AR(1)* model did fit the data well.

7.3.2 Fitting MA(q) Models

Extending the *AR(1)* model, we can add the *MA(q)* process with different orders, such as $q = 1$, 2, and 3, to fit a series of $ARMA(1,q)$ ($q = 1$, 2, 3) as follows:

```
# Fit the MA(1) models and print the model fitting
ar1ma1 = arima(w, order = c(1,0,1));ar1ma1
```

```
## Call: arima(x = w, order = c(1, 0, 1))
##
## Coefficients:
##          ar1    ma1  intercept
##        0.797  0.291      5.255
## s.e.   0.295  0.812      1.203
##
## sigma^2 estimated as 0.935: loglikelihood=-25.7, aic=59.39
```

```
# Fit the MA(2) models and print the model fitting
ar1ma2 = arima(w, order = c(1,0,2));ar1ma2
```

```
## Call: arima(x = w, order = c(1, 0, 2))
##
## Coefficients:
##          ar1    ma1     ma2  intercept
##        0.907  0.312  -0.379      5.241
## s.e.   0.118  0.281   0.262      1.490
##
## sigma^2 estimated as 0.806: loglikelihood=-24.74, aic=59.47
```

```
# Fit the MA(3) models and print the model fitting
ar1ma3 = arima(w, order = c(1,0,3));ar1ma3
```

```
## Call: arima(x = w, order = c(1, 0, 3))
##
## Coefficients:
##          ar1    ma1     ma2    ma3  intercept
##        0.870  0.311  -0.369  0.321      5.268
## s.e.   0.163  0.711   0.543  0.315      1.418
##
## sigma^2 estimated as 0.68: loglikelihood=-24.02, aic=60.03
```

As seen from these model fittings, each model estimated the $ARMA(1,q)$ ($q = 1$, 2, 3) model coefficients along with the standard error (i.e., s.e.). In each model we can see that *ar1* for AR(1) model, as well as *ma1* for *MA(1)*, *ma2* for *MA(2)* model, *ma3* for *MA(3)* model. Similarly, the *intercept* in each model is the estimate of μ. In addition, we can see output for model diagnostics $\hat{\sigma}_\epsilon^2$, along with the *log-likelihood* and the value of *aic*.

For model selection, we compare these added MA processes with the *AR(1)* using the *AIC* as follows:

```
# Model selection based on AIC
AIC(ar1,ar1ma1, ar1ma2,ar1ma3)
```

```
##          df   AIC
## ar1       3 57.61
## ar1ma1    4 59.39
## ar1ma2    5 59.47
## ar1ma3    6 60.03
```

As seen from the values of AICs, the model fitting is getting worse by adding higher-order MA processes and all three MA models have higher *AIC* than the *AR(1)*. This means that the *AR(1)* model is still the best model fitted to this *wage* data. The model diagnostics for *ARMA(1,1)* can be seen in Figure 7.3. As seen in this figure, there is no time series autocorrelation anymore in the residuals, which has similar conclusion as in the *AR(1)* model.

7.3.3 Fitting ARIMA model

To see whether *ARIMA* can fit the data better than the best *AR(1)* model from the above, we can fit a series of *ARIMA(1,d,0)* with $d = 1$, 2, and 3 as follows:

```
# Fit ARIMA(1,d,0) with d = 1
ar1I1 = arima(w, order = c(1,1,0));ar1I1
```

```
##
## Call: arima(x = w, order = c(1, 1, 0))
##
## Coefficients:
##           ar1
##        -0.014
## s.e.    0.236
##
## sigma^2 estimated as 1: loglikelihood=-24.15, aic=52.3
```

```
# Fit ARIMA(1,d,0) with d = 2
ar1I2 = arima(w, order = c(1,2,0));ar1I2
```

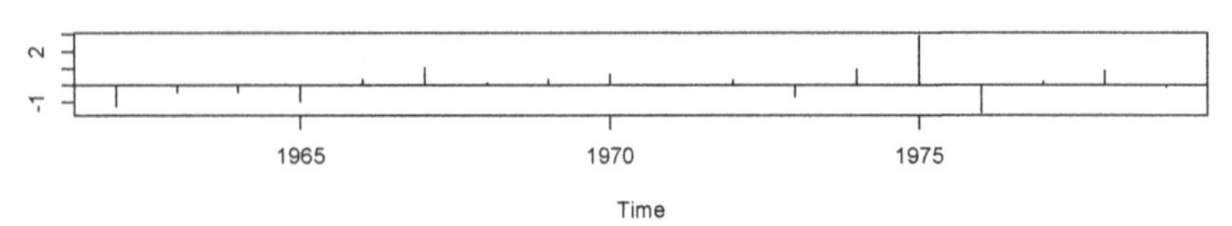

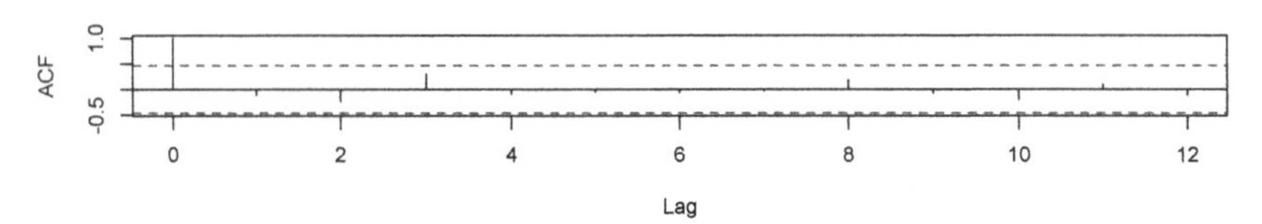

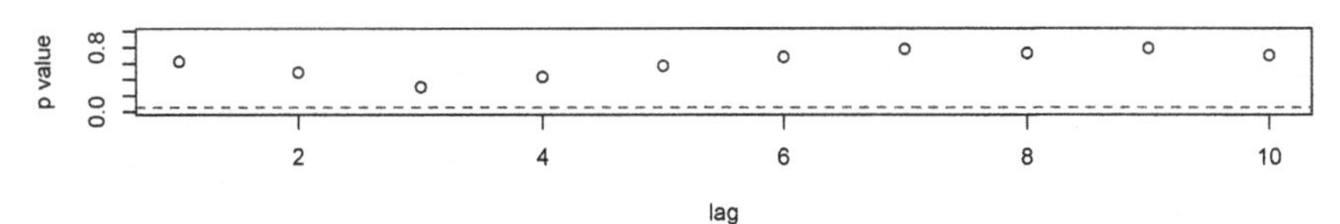

FIGURE 7.3
Model Diagnostics Plot for ARMA(1,1) Model

```
##
## Call: arima(x = w, order = c(1, 2, 0))
##
## Coefficients:
##          ar1
##       -0.337
## s.e.   0.228
##
## sigma^2 estimated as 1.89: loglikelihood=-27.85, aic=59.71
```

```
# Fit ARIMA(1,d,0) with d = 3
ar1I3 = arima(w, order = c(1,3,0));ar1I3
```

```
##
## Call: arima(x = w, order = c(1, 3, 0))
##
## Coefficients:
##          ar1
##       -0.454
```

```
## s.e.    0.223
##
## sigma^2 estimated as 4.8: loglikelihood=-33.16, aic=70.32
```

For model selection, we can again make use of the AICs as follows:

```
AIC(ar1,ar1I1, ar1I2,ar1I3)
```

```
##        df   AIC
## ar1     3 57.61
## ar1I1   2 52.30
## ar1I2   2 59.71
## ar1I3   2 70.32
```

It can be seen that the *ARIMA(1,1,0)* model fits the data the best.

7.3.4 Fitting Time Series Regression

From all the analyses above, the *AR(1)* model and the *ARIMA(1,1,0)* model has the better fit. We can further investigate the regression effects of the independent variables of *consumer price index, unemployment, and minimal wages.* The regression with *ARIMA* model can be used with option *xreg* in *arima* function as follows:

```
# Call xreg for time series regression
ar1reg = arima(w, order = c(1,0,0),
               xreg = cbind(CPI,CPI1, CPI2,u,mw) );
# Print the model fit
ar1reg
```

```
##
## Call: arima(x=w, order=c(1,0,0), xreg=cbind(CPI,CPI1,CPI2,u,mw))
##
## Coefficients:
##          ar1  intercept    CPI   CPI1   CPI2
##        0.438      4.391  0.462  0.167  0.212
## s.e.   0.221      0.465  0.066  0.081  0.044
##            u     mw
##       -0.509  0.042
## s.e.   0.101  0.012
##
## sigma^2 estimated as 0.0968: loglikelihood=-4.63, aic=25.26
```

To see the improvement of model fitting, we can again use AICs to compare this model to the AR(1) model as follows:

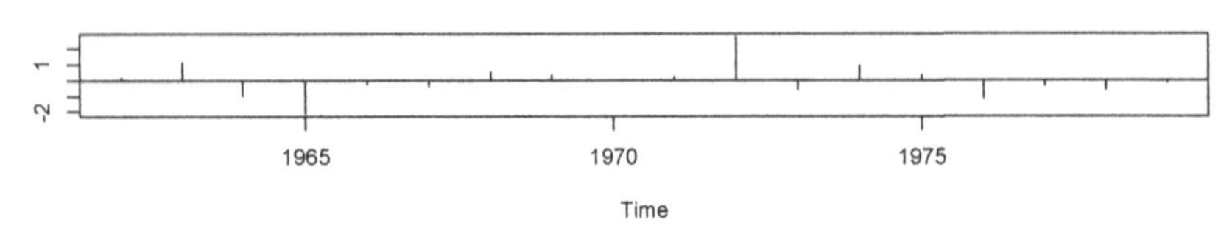

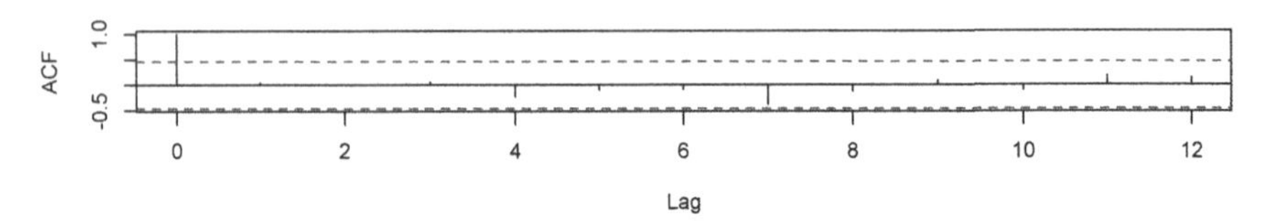

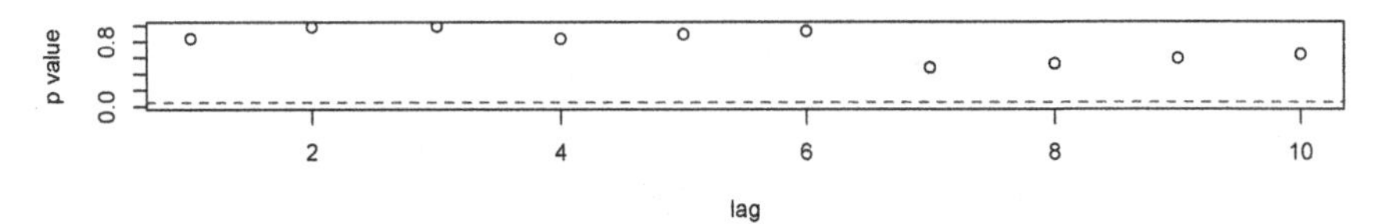

FIGURE 7.4
Model Diagnostics Plot for Regression with AR(1) Residuals

```
AIC(ar1,ar1reg)
```

```
##         df   AIC
## ar1      3 57.61
## ar1reg   8 25.26
```

As seen, the AIC dropped from 57.605 in the *AR(1)* model to 25.262 in the regression with *AR(1)* residuals. This better model fit can be graphically illustrated in Figure 7.4. As seen in this figure, there is no time series autocorrelation anymore in the residuals and the p-values for Box-Ljung tests are getting larger, which indicated better model fit.

With this regression model, we can call *coeftest* to output the statistical inference for estimated time series coefficients.

```
# Call coeftest to output coefficient inference
coeftest(ar1reg)
```

```
##
## z test of coefficients:
```

```
##
##              Estimate Std. Error z value Pr(>|z|)
## ar1           0.4378     0.2206    1.98  0.04723
## intercept     4.3906     0.4654    9.43  < 2e-16
## CPI           0.4619     0.0662    6.98  2.9e-12
## CPI1          0.1669     0.0805    2.07  0.03822
## CPI2          0.2124     0.0435    4.88  1.1e-06
## u            -0.5094     0.1010   -5.04  4.6e-07
## mw            0.0421     0.0123    3.42  0.00063
```

From this model, all the independent variables of *CPI1*, *CPI2*, *CPI3*, *u* and *mw* are now statistically significant. This is different from the conclusion in Table 2.1 from Chapter 2, where only the variables of *CPI*, *CPI3*, and *u* were statistically significant.

It's important to bear in mind that the classical *linear regression* discussed in Chapter 2 operates under the assumption of independent residuals, an assumption frequently breached when dealing with time series data. Conversely, the time series *ARIMA regression* method presented in this chapter addresses this concern by incorporating the inherent time series *autocorrelation* through the utilization of an *ARIMA* process. This approach proves more suitable for the analysis of time series data.

7.4 Monte-Carlo Simulation Validation

As shown in the previous section, we can obtain different conclusions between the *linear regression* and the *ARIMA regression*. An investigation should be conducted on how the *ARIMA regression* is more appropriate than the classical *linear regression* for data with time series autocorrelation. This section is then aimed to design a Monte-Carlo simulation-based study to show the superiority of *ARIMA regression* on parameter estimation and statistical inference if time series autocorrelation is present.

We will show that the classical *linear regression*, which ignores the *autocorrelation* (also known as *serial correlation*) in the residuals can potentially lead to an inflated significance for the regression coefficients. Specifically, when *autocorrelation* is ignored and not accounted for in the analysis, we expect to see the following consequences:

- *Inflated Significance*: The *standard errors* of the *linear regression* coefficients are estimated assuming independence of the residuals. *Autocorrelation* in time series data violates this assumption, potentially leading to *underestimated standard errors*. As a result, *t-statistics* could be higher than they should be, which can lead to *inflated p-values* and a higher likelihood of declaring coefficients as statistically significant when they might not be in reality.

- *Type-I Error*: Ignoring *autocorrelation* can increase the risk of Type-I errors (false positives). The inflated significance levels could lead you to reject null hypotheses for coefficients that shouldn't be rejected.
- *Efficiency Loss*: Ignoring autocorrelation can reduce the efficiency of parameter estimates. *Autocorrelation* introduces additional information into the residuals, and by ignoring it, we're not utilizing all available information, leading to less precise estimates.

To mitigate these issues, the *autoregressive integrated moving-average (ARIMA)* models can address these potential problems caused by *autocorrelation* in linear regression analysis.

7.4.1 Simulation Design

We will design and implement the Monte-Carlo simulation with *R*. Specifically, we will simulate time series data and then fit both the classical *linear regression* model and the time series *ARIMA regression* model. The following *4-step procedure* is designed for this purpose:

- *Step 1: True Parameters in the ARIMA Regression.* We simplify the *ARIMA regression* model to an *AR(1)* model with $K = 1$ in model (7.10). Therefore, the *ARIMA regression* model is $y_t = \beta_0 + \beta_1 X_{1,t} + \beta_2 X_{2,t} + \epsilon_t$ where ϵ_t is an *AR(1)* process with variance σ^2. The true regression parameters are chosen as $(\beta_0, \beta_1, \beta_2)$ = *(b0True, b1True, b2True)* = (1.5, 1.5, 1.5), and the true parameters for the *AR(1)* are σ_ϵ = *sigmaTrue* = 1.5 and the true autocorrelation coefficient $\gamma_1 = 0.5$. The number of observations in this simulation is set to $n = 100$ (interested readers can increase it to larger number of observations, say $n = 1{,}000$, for more accurate results).
- *Step 2: Data Generation.* With the true parameters from *Step 1*, we generate data for the independent variable x_1 from the standard normal distribution using the *R* code *x1 = rnorm(n)*, x_2 from a binomial distributed as *x2 = rbinom(n,size=1, prob=0.5)*, and the autocorrelated error ϵ using *R* function *arima.sim* as *err.sim = arima.sim(model = list(ar = gammaTrue), n = n)*. Then we can construct the *ARIMA regression* data $y_t = \beta_0 + \beta_1 \times x_1 + \beta_2 \times x_2 + \sigma_\epsilon \times \epsilon$ = *b0True + b1True×x1 + b2True×x2+ sigmaTrue×err.sim.*
- *Step 3: Fit the Classical Linear Regression and the ARIMA Regression.* With the data generated from *Step 2*, we can then fit the classical *linear regression* and the *ARIMA regression* to keep track the estimated parameters and standard errors to construct the associated 95% confidence intervals.
- *Step 4: Summarize the Simulation.* To conclude the procedure, *Step 1* through *Step 3* are repetitively executed for a substantial number of iterations, denoted

as *nsimu* (set at, for instance, 1,000 iterations). In this step, we summarize the Monte-Carlo simulations with respect to the parameter estimation and coverage probability to compare the performance between the classical *linear regression* and the *ARIMA regression.*

7.4.2 Simulation Implementation

The above *4-step* procedure can be implemented in *R* as follows with a comprehensive explanations of the aforementioned steps:

```
# Set the random seed for reproducibility
set.seed(333)
#
# Step 1: True Parameters in the ARIMA Regression
#
# Number of observations in the TS
n = 100
# The true AR(1) and regression parameters of (b0,b1,b2)
ParsTrue   = c(0.5,1.5,1.5,1.5)
gammaTrue = ParsTrue[1];
b0True    = ParsTrue[2];
b1True    = ParsTrue[3];
b2True    = ParsTrue[4];
# The true error SD
sigmaTrue = 1.5
#
# The Simulations looping
#
# The number of simulations
nsimu = 1000
# Placehold to keep track of outputs from:
#   lm: 6-columns of parameter,and coverage for b0, b1,b2
#   AR1 model: 8-columns for AR1-gamma, b0,b1,b2
estParMat = matrix(0, nrow=nsimu, ncol=14)
colnames(estParMat) = c("lm.b0.est","lm.b1.est","lm.b2.est",
        "lm.b0.covered","lm.b1.covered","lm.b2.covered",
        "AR1.gamma1.est","AR1.b0.est","AR1.b1.est","AR1.b2.SE",
        "AR1.gamma.covered","AR1.b0.covered",
        "AR1.b1.covered","AR1.b2.covered")
# Start the simulation looping
for(s in 1:nsimu){
#
# Step 2: Data Generation
#
```

```
# Simulate the x variable
x1 = rnorm(n);
x2 = rbinom(n, size=1,prob=0.5)
# Simulate the AR(1)
err.sim = arima.sim(model = list(ar = gammaTrue), n = n)
# Get the data for the dependent variable
y = b0True + b1True *x1 +b2True*x2 + sigmaTrue*err.sim
#
# Step 3: Now Fit the linear and ARIMA regressions
#
# 3.1: Fit a linear model
fit.lm = lm(y~x1+x2)
# Extract the lm fit statistics: summary(fit.lm)
estPar     = summary(fit.lm)$coef[,1]
SE.estPar = summary(fit.lm)$coef[,2]
covered    = (estPar-1.96*SE.estPar < ParsTrue[-1])
 & (estPar+1.96*SE.estPar > ParsTrue[-1])
estParMat[s, 1:3] =  estPar
estParMat[s, 4:6] =  covered
# 3.2: Fit the ARIMA regression
fit.ar1reg = arima(y, order = c(1,0,0),xreg = cbind(x1,x2))
# Extract the arima fit statistics:
# The estimated parameters
estPar     = fit.ar1reg$coef
# The estimated SE for the parameters
SE.estPar = sqrt(diag(fit.ar1reg$var.coef))
# Coverage on whether the true param are within the 95% CI
covered    = (estPar-1.96*SE.estPar < ParsTrue)
 & (estPar+1.96*SE.estPar > ParsTrue)
# Fill the simulation matrix
estParMat[s,7:10]     = estPar
estParMat[s,11:14]   = covered
} # End of s-looping
# Print the dimension of estParMat
dim(estParMat)

## [1] 1000   14

# Print the first 6 simulations
head(estParMat)

##        lm.b0.est lm.b1.est lm.b2.est lm.b0.covered
## [1,]      1.5028     1.327     1.538             1
## [2,]      1.4032     1.289     1.317             1
```

```
## [3,]    0.8042      1.448      1.952            0
## [4,]    1.7140      1.341      1.354            1
## [5,]    1.0821      1.167      1.733            1
## [6,]    1.3559      1.536      1.346            1
##      lm.b1.covered lm.b2.covered AR1.gamma1.est
## [1,]             1             1         0.5779
## [2,]             1             1         0.5168
## [3,]             1             1         0.4339
## [4,]             1             1         0.5869
## [5,]             0             1         0.3945
## [6,]             1             1         0.5683
##      AR1.b0.est AR1.b1.est AR1.b2.SE
## [1,]     1.5894      1.320     1.491
## [2,]     1.2839      1.431     1.523
## [3,]     0.8011      1.358     1.992
## [4,]     1.6127      1.450     1.555
## [5,]     0.9896      1.258     1.865
## [6,]     1.4002      1.711     1.217
##      AR1.gamma.covered AR1.b0.covered
## [1,]                 1              1
## [2,]                 1              1
## [3,]                 1              0
## [4,]                 1              1
## [5,]                 1              1
## [6,]                 1              1
##      AR1.b1.covered AR1.b2.covered
## [1,]              1              1
## [2,]              1              1
## [3,]              1              1
## [4,]              1              1
## [5,]              1              1
## [6,]              1              1
```

It can be seen that we generated a dataframe *estParMat* with 1,000 rows and 14 columns with the first 6 rows printed using the *R* function *head*. The 14 columns are:

- *6-columns for the linear regression*: we keep track of the parameter estimates for β_0, β_1 and β_2 (named as *lm.b0.est,lm.b1.est,lm.b2.est*) and the associated *coverage* on whether the true parameters are covered by the 95% confidence intervals (denoted by *lm.b0.covered,lm.b1.covered,lm.b2.covered*),
- *8-columns for the AR1 regression*: similarly we keep track of the parameter estimates for autocorrelation coefficient γ, β_0, β_1, and β_2 (named as *AR1.gamma1.est, AR1.b0.est, AR1.b1.est, AR1.b2.est*) and the associated *coverage* on whether the true parameters are covered by the 95% confidence

intervals (denoted by *AR1.gamma.covered, AR1.b0.covered, AR1.b1.covered, AR1.b2.covered*).

The Monte-Carlo sampling distributions from these 1,000 simulations for the estimated parameters $(\hat{\beta}_0, \hat{\beta}_1, \hat{\beta}_2, \hat{\gamma}_1)$ from the *AR1 regression* and $(\hat{\beta}_0, \hat{\beta}_1, \hat{\beta}_2)$ from the *linear regression* can be graphically shown in Figure 7.5 using the following *R* code chunk:

```
#
# Step 4: graphical summary
#
par(mfrow=c(2,4))
# Plot the AR1 regression
hist(estParMat[,8],nclass=20, main="",
     xlab="AR1: Beta0")
abline(v=b0True, lwd=3, col="red")
hist(estParMat[,9],nclass=20, main="",
     xlab="AR1: Beta1")
abline(v=b1True, lwd=3, col="red")
hist(estParMat[,10],nclass=20, main="",
     xlab="AR1: Beta2")
abline(v=b2True, lwd=3, col="red")
hist(estParMat[,7], nclass=20, main="",
     xlab="AR1: Gamma1")
abline(v=gammaTrue, lwd=3, col="red")
# Plot the linear regression
hist(estParMat[,1],nclass=20, main="",
     xlab="LM: Beta0")
abline(v=b0True, lwd=3, col="red")
hist(estParMat[,2],nclass=20, main="",
     xlab="LM: Beta1")
abline(v=b1True, lwd=3, col="red")
hist(estParMat[,3],nclass=20, main="",
     xlab="LM: Beta2")
abline(v=b2True, lwd=3, col="red")
```

As seen in Figure 7.5, the estimated parameters are unbiased and all close to the true parameters.

Furthermore from the *estParMat*, we can notice that the columns (3 columns for the *linear regression* and 4 columns for the *AR regression*) for the *coverage* are symbolized by 0 (*not covered*) and 1 (*covered*), which means whether the 95% confidence interval from each regression model *covers* the true parameter. The coverage probability among these 1,000 simulation can be obtained by *averaging* all the 0s and 1s from these 1,000 simulations, which can be done using *R* function *apply* as follow:

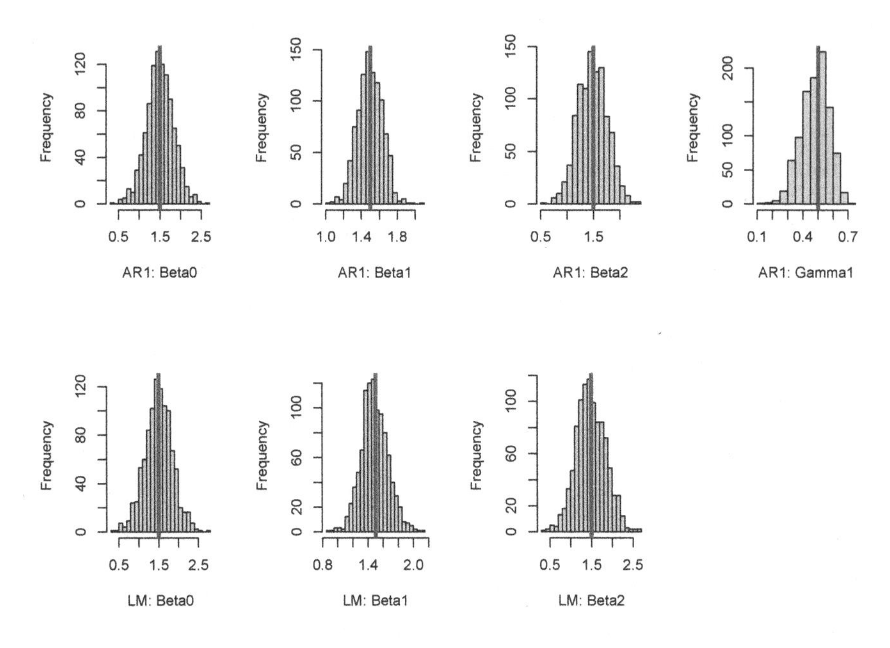

FIGURE 7.5
Monte-Carlo Distributions Overlaid with the True Parameters

```
#
# Step 4: summary the simulations
#
mu = apply(estParMat,2,mean); mu

##          lm.b0.est          lm.b1.est          lm.b2.est
##             1.4996             1.4892             1.4877
##      lm.b0.covered      lm.b1.covered      lm.b2.covered
##             0.8480             0.9440             0.9480
##     AR1.gamma1.est         AR1.b0.est         AR1.b1.est
##             0.4811             1.5040             1.4961
##          AR1.b2.SE AR1.gamma.covered     AR1.b0.covered
##             1.4821             0.9420             0.9390
##     AR1.b1.covered     AR1.b2.covered
##             0.9460             0.9460
```

We can see from these 1,000 simulations that the means for the estimated parameters $(\hat{\beta}_0, \hat{\beta}_1, \hat{\beta}_2)$ are (1.4996414, 1.4892226, 1.4877366) for the *linear regression*, whereas, the means for the parameters $(\hat{\gamma}_1, \hat{\beta}_0, \hat{\beta}_1, \hat{\beta}_2)$ from the *AR1 regression* are (0.4810699,1.5039780, 1.4961354, 1.4821408). It is evident that these parameter estimates are unbiased and all close to the true parameters we used for the 1,000 simulations.

However, the *coverage probabilities* for $(\beta_0, \beta_1, \beta_2)$ in *linear regression* are (0.848, 0.944, 0.948) with the coverage probability for β_0 as 0.848, which is less than the designed 95% confidence probability. In contrast, the coverage probabilities for the *AR1 regression* parameters $(\gamma_1, \beta_0, \beta_1, \beta_2)$ are (0.942, 0.939, 0.946, 0.946) which are all close to the designed 95% confidence probability. Note that these coverage probabilities can be even closer to the desired 95% if we run the simulation more than 1,000.

7.4.3 Summary

This simple Monte-Carlo simulation reveals that the *linear regression* can produce *undercoverage*, where the calculated 95% confidence intervals do not adequately capture the true values of parameters. *Undercoverage* can produce:

- *Inaccurate Statistical Inference*: When the confidence intervals exhibit *undercoverage*, it means they are narrower than they should be. This can lead to inaccurate inference about the true values of parameters and less confidence in the analysis results than initially thought.
- *Increased Risk of Type-I Errors*: *Undercoverage* increases the risk of Type-I errors with higher level of false positives in hypothesis testing. If the confidence intervals are too narrow, it can incorrectly conclude that a parameter is significantly different from a hypothesized value when it's not.

This *undercoverage* can also produce *inflated significance*. In this case, the estimated *standard errors* of the *linear regression* coefficients are *underestimated*. As a result, *t-statistics* are higher than they should be, which can lead to *inflated p-values* and a higher likelihood of declaring coefficients as statistically significant when they might not be in reality. This is very important in financial data analysis.

In summary, we simulated a basic *AR1* (*autoregressive order 1*) process in this section. Interested readers can further expand this Monte-Carlo simulation exercise to other time series scenarios, such as *MA(q)* (*moving-average of order q*) or *ARIMA(p,d,q)* (*autoregressive integrated moving-average of order p, differencing degree d, and moving-average order q*) processes. Exploring these processes can lead to deeper insights and enhanced conclusions to validate the time series modeling whenever the data have time series autocorrelation.

7.5 Discussions

In this chapter, we introduced the analysis of time series. We started the introduction with the time series *autocorrelation function* and *partial autocorrelation function* so to bring the readers' attention to the particular correlation existed in time series data. From there, we then discussed how to model this time series *autocorrelation* with *AR(p)* model, *MA(q)* model, *ARMA(p,q)* model and then the *ARIMA(p,d,q)* model. We further discussed the time series regression with *ARIMA* process to incorporate time-series autocorrelation into regression, and compared it to the classical regression which could give a different conclusion.

7.5.1 Stationary Process

We would like to emphasize that all the *ARIMA* processes are assumed to *stationary* processes. It is easy to understand that we observe a time series with fluctuations from time to previous time points. In *stationary processes*, we assume that the mean, standard deviation, and other statistical properties are similar from one time to the next, and therefore, the *stationary processes* are stochastic probability models for time series with *time-invariant* behavior.

There are two types of *stationary processes*. The first is called *strictly stationary*, where all aspects of its behavior are unchanged by shifts in time. Mathematically, it is defined for every m and n, the distributions of $(y_1, \cdots, y_n)$ and $(y_{1+m}, \cdots, y_{n+m})$ are the same. In this *strictly stationary*, the probability distribution of a sequence of n observations does not depend on their time origin. *Strict stationarity* is a very strong assumption due to the requirement that *all aspects* of stochastic behavior be constant in time. Often, this *strictly stationary* can be relaxed to the *weak stationarity*. A process is *weakly stationary* as long as its mean, variance, and covariance are unchanged by time, which can be mathematically specified with requirements:

- $E(y_t) = \mu$ for all time t;
- $Var(y_t) = \sigma^2$ for all t; and
- $Cov(y_t, y_s) = \gamma(|t - s|)$ for all t and s for some function $\gamma(h)$.

As seen from this definition, the mean and variance are constant with time and the covariance between any two observations at time t and s depends only on the lag of $t - s$. For example, if the process is *weakly stationary*, then the covariance between y_5 and y_{15} is the same as the covariance between y_9 and y_{19}, since each pair is separated by ten units of time (i.e., 15-5 = 19-9 = 10). In this sense, the *weakly stationary* only requires that the means, variance, and covariances, not other distributional characteristics such as quantiles, skewness, and kurtosis, are stationary. Thus, the *weakly stationary* is sometimes referred to as *covariance stationary*. The function $\gamma(h)$ is the *autocovariance function* which is used to describe the ACF in the time series.

Stationary time series can be easily modeled as shown in the ARIMA model presented in this chapter. For some of the *non-stationary* processes, the *difference operator* in *ARIMA* can also make it to *stationary process*. Therefore the ARIMA model discussed in this chapter can be used to model a large class of time series.

Most importantly in the *ARIMA(p,d,q)* process, we need to specifically mention two special cases, which are the *white noise* process and the *random walk* process.

7.5.2 White Noise Process is a Stationary Process

An *ARIMA(0, 0, 0)* model becomes a *white noise* process of y_t or more commonly denoted by ϵ_t. Using *integrated* notation, the *white noise* process is in fact a *I(0)* process and is typically defined as follows:

$$\begin{aligned} E(\epsilon_t) &= 0 \\ E(\epsilon_t^2) &= \sigma^2 \\ E(\epsilon_t \epsilon_{t'}) &= 0 \text{ for } t \neq t'. \end{aligned} \tag{7.11}$$

If necessary, we can further assume that ϵ_t in equation (7.11) to follow a *normal* distribution as $\epsilon_t \sim N(0, \sigma^2)$, then we would have a *normal white noise* process or *Gaussian white noise* process. A *random noise* process is always a *stationary* process.

7.5.3 Random Walk Process is a Non-Stationary Process

A *random walk* is an *ARIMA(0, 1, 0)* model defined as follows:

$$y_t = y_{t-1} + \epsilon_t = y_0 + \sum_{i=1}^{t} \epsilon_i \tag{7.12}$$

where ϵ_t is the *white noise* process defined in equation (7.11). Using *integrated* notation, the *random walk* process is in fact a $I(1)$ process.

It can be easily shown that $E(y_t) = y_0$ and $Var(y_t) = t\sigma^2$. Therefore, *random walk* is not a *stationary* process since its variance is not a constant and increases linearly with the time t. Even though the *random walk* y_t itself is not *stationary* process, its *difference(integrated)* process is in fact a *stationary* process since $\Delta y_t = \epsilon_t$, which is the *random noise* process.

Both the *white noise* $I(0)$ and *random walk* $I(1)$ will be the foundational concepts for *cointegration* to be discussed in Chapter 9

7.5.4 Stationary to Heteroskedasticity

Most real financial time series do not exhibit *stationarity*, one possible solution is to apply a transformation (such as log-transformation) to make them approximately stationary. Another option is to directly model *non-stationarity* with an advanced time series model, such as the *Auto-Regressive Conditional Heteroskedasticity (ARCH)* or the *Generalized Auto-Regressive Conditional*

Heteroskedasticity (GARCH). In these *ARCH/GARCH models*, we model the conditional variance with a structure very similar to the structure of the *conditional expectation* in an *ARMA* model and we will discuss these models in Chapter 8.

7.6 Exercises

Section 7.4 was designed to show the differences between the general *ARIMA* regression and the classical *linear regression* if the data are time series correlated. For further understanding, let's consider a simplified version to investigate the long-term *mean* of the time series.

1. Simulate an *AR(1)* process with the autocorrelation $\gamma = 0.1$ and 1,000 observation, i.e., $y_t(t = 1, \cdots, n)$ with $n = 1000$.

   Hint: Using the following *R* code chunk:

```
# Use "arima.sim" to generate the correlated data
y = arima.sim(model = list(ar = 0.1), n=1000)
```

2. Calculate the long-term *mean* using *R* function *mean.*

   Hint: Using the following *R* code chunk:

```
# Calculate the data mean
m.y = mean(y)
# Print the mean
m.y
```

3. Calculate the long-term *mean* using *lm*. Is this *mean* equal to the *mean* from *Exercise #2*? Why?

   Hint: Using the following *R* code chunk:

```
# Fit "lm"
fit.lm = lm(y~1)
# Look at the summary
summary(fit.lm)
# Output the "mean" estimate from "lm"
m.lm = coef(fit.lm);m.lm
```

4. Calculate the *mean* using *arima* model. Is the estimated *mean* from *AR(1)* model equal to the *mean* from *Exercise #2*? Why?

Hint: Using the following *R* code chunk:

```
# Fit "AR(1)"
fit.AR1 = arima(y, order = c(1,0,0))
# Print the model fit
fit.AR1
# Extract the estimated "mean"
m.AR1 = coef(fit.AR1)[2]; m.AR1
```

5. Change $\gamma = 0.1$ to $\gamma = 0.9$, redo *Exercise #1* to *Exercise #4*. what do you find?

8

Generalized AutoRegressive Conditional Heteroskedasticity Model

Financial markets often exhibit a striking tendency known as volatility clustering, where periods of high volatility are followed by more periods of high volatility, and the same holds true for low-volatility periods. Understanding and effectively modeling time-varying volatility is paramount in the world of finance.

To navigate these market dynamics, financial analysts rely on a range of tools and models. We usually begin by analyzing historical market data, calculating various measures of historical volatility over different timeframes. Moving averages are typically employed to smooth price data and identify trends in volatility, while classical statistical models like the *ARIMA* models discussed in Chapter 7 are very useful in modeling the conditional expectation of a process given the past, if the conditional variance given the past is constant. However, most typically, the conditional variance given the past will not be *nonconstant.* This is where the more advanced *Autoregressive Conditional Heteroskedasticity (ARCH)* and *Generalized Autoregressive Conditional Heteroskedasticity (GARCH)* models become useful to estimate how past volatility influences future volatility, capturing the clustering effect in the dynamic world of financial markets.

The terms *Time Series ARCH Model* and *GARCH Model* are often used interchangeably or in similar contexts because they both deal with modeling and forecasting volatility in time series data. However, there are some differences between the two.

Historically, the *ARCH model* was introduced by Robert Engle in 1982 (Engle, 1982). It is the foundational model in this family and primarily focuses on modeling the *conditional variance* of a time series, meaning it models how the variance of the series changes over time based on past observations. The *GARCH model*, however, was introduced by Tim Bollerslev in 1986 (Bollerslev, 1986). It is an extension of the *ARCH model* and offers more flexibility by allowing lagged values of both the *conditional mean* and the *conditional variance* in the model. The *GARCH model* is then more versatile and widely used because it can capture a broader range of volatility patterns.

DOI: 10.1201/9781003469704-8

In *model components*, the *ARCH model* typically consists of an *autoregressive (AR)* component for modeling the *conditional mean* of the time series and a separate *autoregressive component* for modeling the *conditional variance*. It is often represented as *ARCH(pV)*, where pV represents the order of the *autoregressive conditional variance* component (to distinguish between the *AR(p)* for the *autoregressive* conditional mean with order p). The *GARCH model* includes the same *autoregressive (AR)* component for modeling the *conditional mean* as the *ARCH model.* In addition to this, it incorporates an *autoregressive* component for modeling the *conditional variance* and a *moving average (MA)* component for modeling the *conditional variance* as well. The *GARCH model* is usually represented as *GARCH(pV, qV)*, where pV represents the order of the *autoregressive conditional variance* component, and qV represents the order of the *moving average conditional variance* component.

Therefore, the *ARCH models* are less flexible than *GARCH models* because they only capture the *conditional variance changes* without considering the *conditional mean changes* in a time series. The *GARCH models* are more flexible because they can capture both the *conditional mean and conditional variance changes.* This added flexibility often leads to better model performance, where both mean and volatility dynamics are essential.

To introduce the *ARCH model* and *GARCH model* in this chapter, we will make use of the compiled historical *VIX(Volatility Index)* index daily data from January 2nd, 1990 to June 27th, 2023 to illustrate the *GARCH* modeling using *R* over the classical *ARIMA* modeling described in Chapter 7.

8.1 The Volatility Index Data

8.1.1 The Volatility Index Briefly Explained

The *VIX* is often referred to as the *fear gauge* or *fear index* in financial markets, which is a crucial measure of expected market volatility, specifically for the U.S. equity market. The *VIX* index is based on the S&P 500 Index (*SPX*), one of the most important benchmarks for U.S. stocks.

The *VIX* is calculated by the Chicago Board Options Exchange (CBOE) using options prices on the S&P 500 Index. It represents the market's consensus on future volatility. Specifically, it aggregates the weighted prices of a wide range of *SPX* put and call options with varying strike prices and expiration dates. A higher *VIX* value indicates that market participants expect greater price swings or increased volatility in the S&P 500 over the next 30 days. This is often a sign of fear or uncertainty in the market. Investors may view a high *VIX* as a signal to exercise caution and consider risk mitigation strategies.

The *VIX* is not a direct predictor of market direction (up or down); rather, it measures the expected magnitude of price swings for *market volatility*. High *VIX* values don't necessarily mean markets will decline, but they suggest greater uncertainty and potential for larger price fluctuations. Therefore, many traders and investors use the *VIX* as a tool for risk management. For example, they may adjust their portfolio allocations or hedge their positions when the *VIX* rises significantly to protect against potential market downturns. Generally speaking, the *VIX* is a critical gauge of expected market volatility, and its interpretation can be a valuable tool for market participants.

8.1.2 The VIX Data

Daily data from January 2nd, 1990 to June 27th, 2023 (except weekends) are compiled in the writing of this chapter. There are 8,441 observations with 5 columns (i.e., variables) of *OPEN*, *HIGH*, *LOW*, and *CLOSE*, which is saved in *VIX_history.csv* file, which can be loaded in *R* as follows:

```
# Read the data into R
dVIX = read.csv("data/VIX_history.csv", header=T)
# Create the Date variable
dVIX$Date = as.Date(dVIX$DATE, format = "%m/%d/%Y")
# Print the dimension of the data
dim(dVIX)
```

```
## [1] 8441    6
```

```
# Print the first 6 observations
head(dVIX)
```

```
##        DATE  OPEN  HIGH   LOW CLOSE       Date
## 1 1/2/1990 17.24 17.24 17.24 17.24 1990-01-02
## 2 1/3/1990 18.19 18.19 18.19 18.19 1990-01-03
## 3 1/4/1990 19.22 19.22 19.22 19.22 1990-01-04
## 4 1/5/1990 20.11 20.11 20.11 20.11 1990-01-05
## 5 1/8/1990 20.26 20.26 20.26 20.26 1990-01-08
## 6 1/9/1990 22.20 22.20 22.20 22.20 1990-01-09
```

Note that the data for the original *DATE* is in character *month*, *date*, and *year* format with *year* in 4-digit format (such as *1/2/1990* in the first observation). This variable should be converted to the *R Date* format using *R* function *as.Date* with the option *format = "%m/%d/%Y"*. Notice that the capital *Y* is used due to the *year* is in 4-digit format. Otherwise, lower case *y* should be used if the *year* is in 2-digit format, such as *1/2/90*.

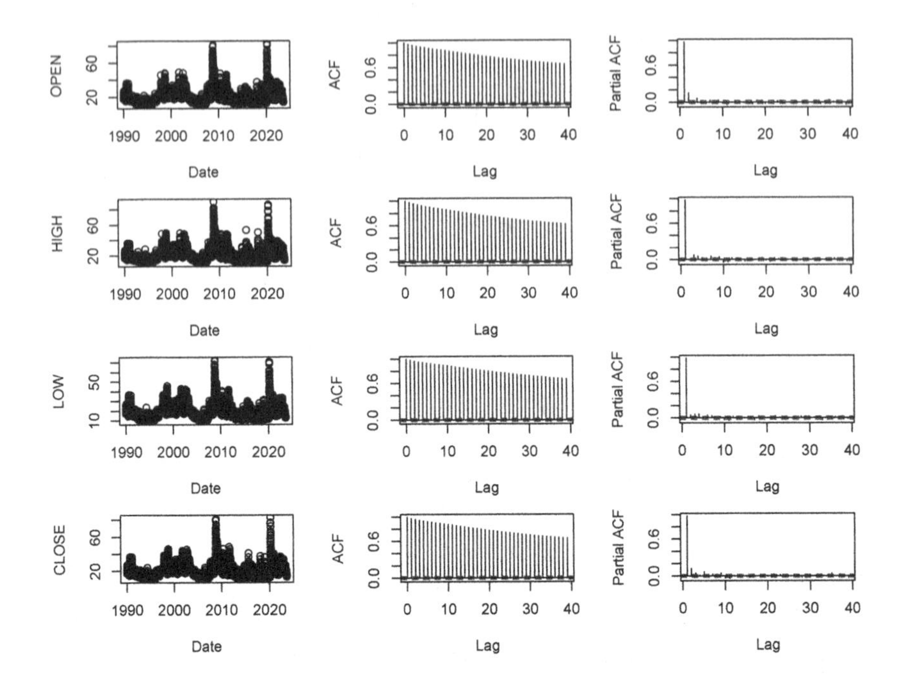

FIGURE 8.1
VIX Time Series Plot

8.1.3 Descriptive Analysis

In this chapter, we will only use the daily *OPEN* data to illustrate the *GARCH* modeling. The other 3 *VIX* time series will be left as exercises and also be used in Chapter 9 for *Cointegration.*

Let's start with the time series (with 8,441 observations) plot in the descriptive data analysis to examine the time series trend and volatility. This can be easily done using *R* function *plot* as follows:

```
# Plot the VIX time series
plot(dVIX$OPEN,type="l", ylab="OPEN VIX", xlab="Time")
```

From Figure 8.1, we can observe distinct patterns in the *OPEN VIX* (*CBOE Volatility Index*) time series data. Beginning on January 2nd, 1990, represented as *Time* = 1, the VIX exhibits notable fluctuations.

Initially, the VIX starts around the 20 mark but undergoes a decline, reaching approximately 15 by the close of 1993 (*Time* = 1,000). Subsequently, it returns to 20 by the conclusion of 1996 (*Time* = 1770). Following this period, an extended phase emerges where the VIX remains constantly about 30. This phase spans from the end of 1996 and extends until the close of March 2003 ($Time = 3337$).

Notably, there are intervals of heightened *VIX* values, marking periods of increased market volatility. These include the period from mid-September 2008 (specifically, at $Time = 4712$ on September 15, 2008) to mid-July 2010 (reaching $Time = 5174$ on July 16, 2010). Another such period occurs from mid-February 2020 (at $Time = 7587$ on February 18, 2020) and continues until late 2022 (reaching $Time = 8270$ on October 26, 2022). These spikes in the *VIX* are often associated with significant economic events, such as the global financial crisis and the COVID-19 pandemic.

To investigate the time series correlation, we can make use of the ACF/PACF plots introduced and used in Chapter 7 to visualize the time series autocorrelation as in Figure 8.2 with the following *R* code chunk:

```
par(mfrow=c(1,2),mar=c(4,4,1,2),oma=c(1,1,1,1) )
# Extract the OPEN VIX data and rename it
#      as "vix" for simplicity
vix=dVIX$OPEN
# Call acf to plot the autocorrlation for VIX
acf(vix,xlab="ACF for OPEN VIX", ylab="ACF", main="")
# Call pacf to plot the partial autocorrlation for VIX
pacf(vix,xlab="PACF for OPEN VIX", ylab="PACF", main="")
```

As seen in Figure 8.2, there exists at least order-2 autocorrelation. We can further compute the Ljung–Box (Ljung and Box, 1978) test statistic to statistically test the null hypothesis of independence using *R* function *Box.test* to the *lag* of 10 as follows:

```
# Call Box.test to use Ljung-Box test to test independence
Box.test(vix, lag = 10, type = "Ljung")
```

```
##
##  Box-Ljung test
##
## data:  vix
## X-squared = 72166, df = 10, p-value <2e-16
```

This *Ljung-Box test* indicates that there is a statistically significant time series autocorrelation with p-value < 0.0001, and further analysis should be done to model and incorporate this *VIX* time series autocorrelation.

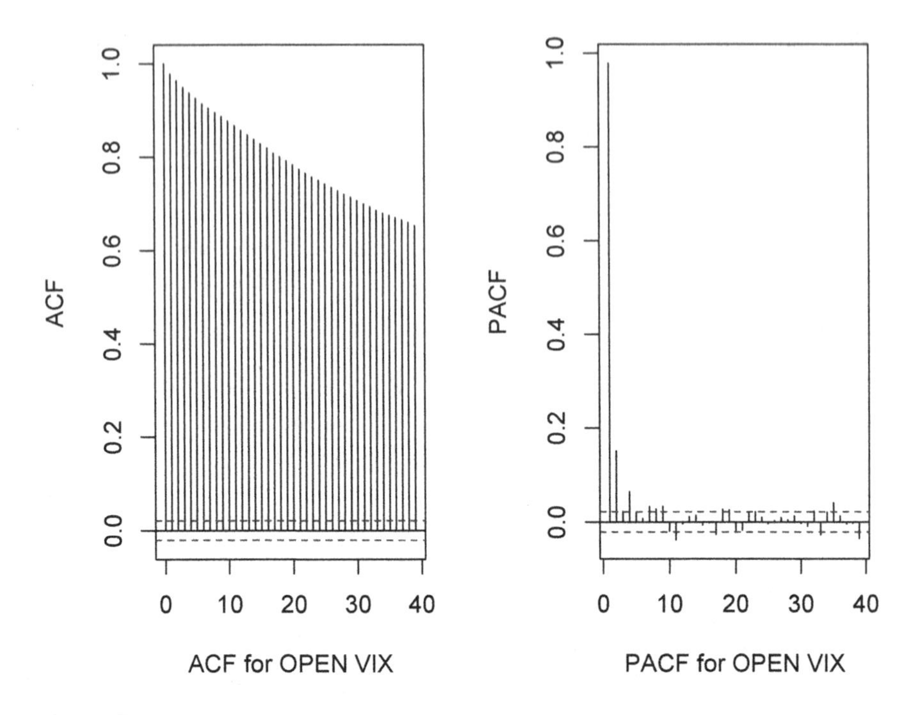

FIGURE 8.2
ACF and PACF for OPEN VIX Data

8.1.4 ARIMA Modeling

Based on the above preliminary analysis, we can now proceed to model *VIX* with *ARIMA* models. To simplify this process, we fit a series of *ARIMA(p,d,q)* models with lower-order of $d = 0$, and p and q not exceeding 2 to keep this fitting simpler so that we will go further to *GARCH* modeling to catch the volatility pattern. Interested readers can consider the higher-order *ARIMA(p,d,q)* models with larger p and q values to capture the time series pattern.

We can fit four *ARIMA* models as implemented in the following *R* code chunk:

```
# Fit the AR1 model
ar1    = arima(vix, order = c(1, 0, 0))
ar1
```

```
##
## Call: arima(x = vix, order = c(1, 0, 0))
##
## Coefficients:
##          ar1  intercept
```

```
##        0.978       19.74
## s.e.   0.002        0.83
##
## sigma^2 estimated as 2.73: loglikelihood=-16214, aic=32435
```

```
# Fit the AR2 model
ar2     = arima(vix, order = c(2, 0, 0))
ar2
```

```
##
## Call: arima(x = vix, order = c(2, 0, 0))
##
## Coefficients:
##          ar1    ar2  intercept
##        0.831  0.151     19.737
## s.e.   0.011  0.011      0.964
##
## sigma^2 estimated as 2.67: loglikelihood=-16117, aic=32242
```

```
# Fit the AR2MA1 model
ar2ma1 = arima(vix, order = c(2, 0, 1))
ar2ma1
```

```
##
## Call: arima(x = vix, order = c(2, 0, 1))
##
## Coefficients:
##          ar1     ar2     ma1  intercept
##        1.342  -0.350  -0.510      19.73
## s.e.   0.084   0.082   0.078       1.09
##
## sigma^2 estimated as 2.66: loglikelihood=-16107, aic=32224
```

```
# Fit the AR2MA2 model
ar2ma2 = arima(vix, order = c(2, 0, 2))
ar2ma2
```

```
##
## Call: arima(x = vix, order = c(2, 0, 2))
##
## Coefficients:
##          ar1    ar2    ma1     ma2  intercept
##        0.306  0.669  0.516  -0.099     19.735
## s.e.   0.194  0.192  0.195   0.035      0.989
```

```
##
## sigma^2 estimated as 2.66: loglikelihood=-16112, aic=32235
```

```
# Output the AICs for model selection
AIC(ar1,ar2,ar2ma1,ar2ma2)
```

```
##         df   AIC
## ar1      3 32435
## ar2      4 32242
## ar2ma1   5 32224
## ar2ma2   6 32235
```

Among the four *ARIMA* models, the *ARIMA(2,0,1)* (denoted by *ar2ma1*) has the lowest AIC value of 32,224.38 and will be used for further analysis.

With this model *ar2ma1*, we can extract the residuals (denoted by *resid*) from this fitted model to examine the residual time series pattern using the ACF/PACF. Additionally, we examine the variance pattern as characterized by the second-order residuals (denoted by *resid^2*) as follows:

```
# Get the residuals from ARMA(2,1)
resid = ar2ma1$residuals
# Figure layout
par(mfrow=c(2,2),mar=c(4,4,1,2),oma=c(1,1,1,1) )
# acf to plot the autocorrlation for residuals from AR2MA1
acf(resid,xlab="ACF for Residuals from AR2MA1",
    ylab="ACF", main="")
# pacf to plot the partial autocorrlation for VIX
pacf(resid,xlab="PACF for Residuals from AR2MA1",
     ylab="PACF", main="")
# acf to plot the autocorrlation for squared VIX
acf(resid^2,xlab="ACF for Squared-Residuals from AR2MA1",
    ylab="ACF", main="")
# pacf to plot the partial autocorrlation for squared-VIX
pacf(resid^2,xlab="PACF for Squared-Residuals from AR2MA1",
     ylab="PACF", main="")
```

As seen from Figure 8.3, the *ARMA(2,1)* model captures the time series autocorrelations very satisfactorily as shown in the two plots in the first row. This is in contrast to the ACF/PACF in Figure 8.2. However, there are still significant second-order autocorrelations as shown in the two plots in the second row, which shows how past volatility influences future volatility. This is exactly how the *GARCH* models come to play to model the conditional variance component with the *autoregressive (AR)* and *moving average (MA)* components.

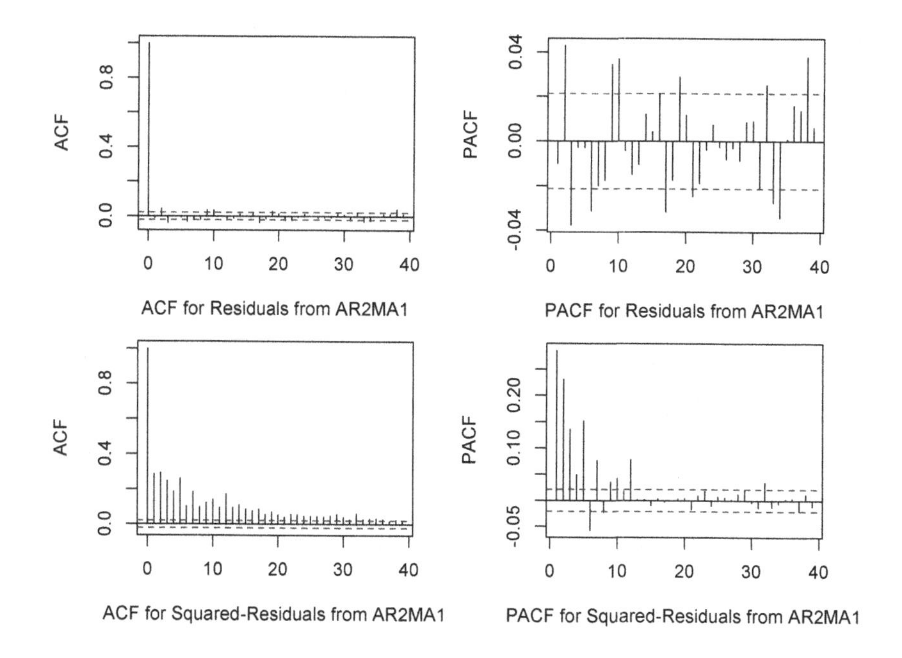

FIGURE 8.3
ACF and PACF for the Residuals from AR2MA1 Model

8.2 The GARCH Modeling

8.2.1 From General Regression to *ARCH* Models

The general form of regression is

$$y_t = f(X_{1,t}, \ldots, X_{K,t}) + \sigma\epsilon_t, \tag{8.1}$$

where the function f(·) is the conditional expectation of y_t given the K regression variables of $X_{1,t}, \ldots, X_{K,t}$ where $t = 1, \cdots, N$. The error term ϵ_t is typically assumed to be independent with an expectation 0 and a constant conditional variance of 1 so that the variance of y_t can be a constant as $var(y_t) = \sigma^2$.

However, the constant conditional variance σ^2 is not realistic in real financial data. Therefore, we should modify this general regression model in equation (8.1) to account for conditional *heteroskedasticity* in financial time series data as follows:

$$y_t = f(X_{1,t}, \ldots, X_{K,t}) + \sigma(X_{1,t}, \ldots, X_{K,t})\epsilon_t \tag{8.2}$$

where $\sigma(X_{1,t}, \ldots, X_{K,t})$ is the conditional *variance function* of y_t given the K regression covariates of $X_{1,t}, \ldots, X_{K,t}$, which should be modeled based on the characteristics of financial data. The *GARCH* models are then developed to model this *variance function*.

For simplicity, we rewrite $\sigma(X_{1,t}, \ldots, X_{K,t})$ as σ_t in equation (8.2) which should be thought to be linked to the regression covariates of $X_{1,t}, \ldots, X_{K,t}$ at time t. With this notation, the basic *ARCH(1)* model can be defined as

$$y_t = \epsilon_t \sqrt{\omega + \alpha y_{t-1}^2} \tag{8.3}$$

as the very first special case in equation (8.2) where the f is equal to 0 and $\sigma_t = \sqrt{\omega + \alpha y_{t-1}^2}$.

If we square both sides of the *ARCH(1)* model in equation (8.2), we can see that the *ARCH(1)* is as follows:

$$y_t^2 = \omega + \alpha y_{t-1}^2 \tag{8.4}$$

which is in fact an *AR(1)* model in y_t^2 (not to y_t). This is exactly we plotted the squared residuals in Figure 8.3 to examine the ACF/PACF for σ_t^2, as denoted by *resid^2* in Figure 8.3.

However, as seen in Figure 8.3 (in the second row), the ACFs and PACFs exhibit the autocorrelations with more than order 1 and therefore the *ARCH(1)* model above should be extended to *ARCH(pV)* to model the higher-order autoregressive conditional heteroskedasticity (ARCH) with order $pV > 1$. The general *ARCH(pV)* model with order pV is defined as extension of the *ARCH(1)* as follows model:

$$y_t = \epsilon_t \sqrt{\omega + \alpha_1 y_{t-1}^2 + \cdots + \alpha_{pV} y_{t-pV}^2} \tag{8.5}$$

where $\sigma_t = \sqrt{\omega + \alpha_1 y_{t-1}^2 + \cdots + \alpha_{pV} y_{t-pV}^2}$ and the *ARCH(pV)* parameter of $\alpha_1, \cdots, \alpha_{pV}$ are to be estimated with the theory of maximum likelihood estimation.

In summary, the *Autoregressive Conditional Heteroskedasticity (ARCH)* is designed to capture the volatility often observed in financial data, where periods of high volatility tend to be followed by more high volatility and vice versa as depicted in Figure 8.1. Here are some brief explanations of the *ARCH* model to be attention to based on the above mathematical formulations:

- *Conditional Volatility*: *ARCH* models focus on modeling the conditional variance (volatility) of a time series. In other words, it aims to capture how the variance of the series changes over time given past observations.

- *Autoregressive Component*: *ARCH* models typically include an autoregressive component. This means that the current conditional variance is modeled as a function of past conditional variances. This captures the idea that past volatility affects current volatility.
- *White Noise Error Term*: In an *ARCH* model, the *error term* (also called *innovation*) is assumed to be a *white noise* process, meaning it has constant variance and is uncorrelated over time.
- *Parameters Estimation*: The parameters of an *ARCH* model are estimated using methods of *maximum likelihood estimation (MLE)*. These parameters determine how past conditional variances impact the current conditional variance.

While *ARCH* models are essential, generalized versions called *GARCH (Generalized Autoregressive Conditional Heteroskedasticity)* models are more commonly used. *GARCH* models include both *autoregressive* and *moving average* components in modeling conditional variance, making them more flexible and effective in capturing volatility patterns.

8.2.2 From *ARCH* to *GARCH*

Extending the *ARCH* model to a *GARCH* model to capture different volatility patterns, *GARCH* models include both *autoregressive (AR)* and *moving average (MA)* components in modeling conditional variance. The *ARCH(pV)* model in equation (8.5) can be extended to include the *MA(qV)* component to have a *GARCH(pV, qV)* model as follows:

$$\sigma_t^2 = \omega + \sum_{i=1}^{pV} \alpha_i y_{t-i}^2 + \sum_{j=1}^{qV} \beta_j \sigma_{t-j}^2. \tag{8.6}$$

In this equation, ω represents a constant term, just as in the *ARCH* model, $\alpha_i (i = 1, \cdots, pV)$ are *ARCH* parameters that capture the impact of past squared observations, and $\beta_j (j = 1, \cdots, qV)$ are GARCH parameters that capture the impact of past conditional variances.

As seen from this equation, the *GARCH* model allows us to account for not only the influence of past squared observations but also for the influence of past conditional variances on the current conditional variance. This added flexibility makes *GARCH* models more capable of capturing various types of volatility patterns, including long-term persistence and short-term clustering.

In this *GARCH* model, the estimated parameters (i.e., the αs and βs) determine how past observations and past conditional variances affect the current conditional variance in the time series.

8.2.3 R Packages

In *R*, there are several packages to be used to fit *GARCH* models to financial time series data. We list here a few commonly used packages. Interested readers can search *R* community for more packages.

- *rugarch*: The *rugarch* package is a widely used package for univariate *GARCH* modeling in *R* maintained by *Alexios Galanos* (alexios@4dscape.com). It provides a comprehensive set of functions and tools for estimating various *GARCH* models, including *GARCH, GJR-GARCH*, and *EGARCH.* It also offers options for different distributions of the error term. Here are the sample *R* code to install the package, load the package to *R*, and for help on this package:

```
# Install the package
install.packages("rugarch")
# Load the R package to R
library(rugarch)
# For help of this package
library(help = rugarch)
```

- *fGarch*: The *fGarch* package is another popular choice for *GARCH* modeling to analyze and model heteroskedastic behavior in financial time series, which is maintained by *Georgi N. Boshnakov* (georgi.boshnakov@manchester.ac.uk). This package offers a user-friendly interface for estimating *GARCH* models and provides various diagnostic tools and visualization functions. We will use this package to analyze the *VIX* data. Here are the sample *R* code to install the package, load the package to *R*, and for help on this package:

```
# Install the package
install.packages("fGarch")
# Load the package into R
library(fGarch)
# For help on this package
library(help=fGarch)
```

- *tseries*: The *tseries* package is primarily focused on time series analysis and includes functions for *GARCH* modeling. While it may not be as feature-rich as *rugarch* or *fGarch*, it can be a good choice for basic *GARCH* modeling. Here are the sample *R* code to install the package, load the package to *R*, and for help on this package:

```
# Install the package
install.packages("tseries")
# Load the package into R
library(tseries)
```

```
# For help on this package
library(help=tseries)
```

- *mgarch*: The *mgarch* package is specifically designed for *multivariate GARCH* modeling, allowing you to model the conditional variances and covariances of multiple time series simultaneously. Here are the sample *R* code to install the package, load the package to *R* and for help on this package:

```
# Install the package
install.packages("mgarch")
# Load the package into R
library(mgarch)
# For help on this package
library(help=mgarch)
```

- *rmgarch*: Similar to *mgarch*, the *rmgarch* package is designed for *multivariate GARCH* modeling but provides more advanced features and options for modeling and forecasting. Here are the sample *R* code to install the package, load the package to *R* and for help on this package:

```
# Install the package
install.packages("rmgarch")
# Load the package into R
library(rmgarch)
# For help on this package
library(help=rmgarch)
```

These packages offer various *GARCH* model specifications, diagnostic tools, and visualization capabilities to help you analyze and forecast financial time series data. Depending on your specific modeling needs and preferences, you can choose the package that best suits your requirements.

8.3 Data Analysis with GARCH Modeling

We make use of the *fGarch* package to analyze the *VIX* data. We first load the package as follows:

```
# Load the fGarch package
library(fGarch)
```

8.3.1 Model Fitting

With this package, we can then fit a series of *GARCH* models to improve the *ARIMA* (i.e., *ARIMA(2,0,1)*) model fitting in Section 8.1.4. Specifically, we will fit the following 12 models:

- ARMA(2,1) + GARCH(1,0)
- ARMA(2,1) + GARCH(2,0)
- ARMA(2,1) + GARCH(3,0)
- ARMA(2,1) + GARCH(1,1)
- ARMA(2,1) + GARCH(2,1)
- ARMA(2,1) + GARCH(3,1)
- ARMA(2,1) + GARCH(1,2)
- ARMA(2,1) + GARCH(2,2)
- ARMA(2,1) + GARCH(3,2)
- ARMA(2,1) + GARCH(1,3)
- ARMA(2,1) + GARCH(2,3)
- ARMA(2,1) + GARCH(3,3)

Notice that in the *R* implementation, we use *control=list(maxit=10000)* to change the max iterations from the default of 100 to 10,000 in the maximum likelihood estimation to improve the probability of convergence.

The list of the above models can be fitted using the following *R* code chunk:

```
# Fit ARMA(2,1) + GARCH(1,0)
ar2ma1garch10 <- garchFit(~ arma(2,1) + garch(1, 0),
                    data=vix, trace=F, control=list(maxit=10000))
summary(ar2ma1garch10)
```

```
##
## Title:
##  GARCH Modelling
##
## Call: garchFit(formula = ~arma(2, 1) + garch(1, 0),
##  data = vix, trace = F, control = list(maxit = 10000))
##
## Mean and Variance Equation:
##  data ~ arma(2, 1) + garch(1, 0)
##
## Conditional Distribution:
```

```
##  norm
##
## Coefficient(s):
##       mu      ar1      ar2      ma1    omega   alpha1
## 0.90187  0.16011  0.78857  0.82039  1.33898  0.55321
##
## Std. Errors:
##  based on Hessian
##
## Error Analysis:
##         Estimate  Std. Error  t value Pr(>|t|)
## mu       0.90187     0.07990   11.287   <2e-16 ***
## ar1      0.16011     0.07145    2.241    0.025 *
## ar2      0.78857     0.06911   11.411   <2e-16 ***
## ma1      0.82039     0.06997   11.725   <2e-16 ***
## omega    1.33898     0.02867   46.710   <2e-16 ***
## alpha1   0.55321     0.02583   21.422   <2e-16 ***
## ---
## Log Likelihood:
##  -14936    normalized:  -1.769
##
## Standardized Residuals Tests:
##                                 Statistic p-Value
##  Jarque-Bera Test   R    Chi^2  19657     0
##  Shapiro-Wilk Test  R    W      NA        NA
##  Ljung-Box Test     R    Q(10)  78.12     1.173e-12
##  Ljung-Box Test     R    Q(15)  95.01     1.142e-13
##  Ljung-Box Test     R    Q(20)  115.7     1.776e-15
##  Ljung-Box Test     R^2  Q(10)  654.4     0
##  Ljung-Box Test     R^2  Q(15)  785.1     0
##  Ljung-Box Test     R^2  Q(20)  934       0
##  LM Arch Test       R    TR^2   471.9     0
##
## Information Criterion Statistics:
##   AIC   BIC   SIC  HQIC
## 3.540 3.545 3.540 3.542
```

```
# Fit ARMA(2,1) + GARCH(2,0)
ar2ma1garch20 <- garchFit(~ arma(2,1) + garch(2, 0),
                   data=vix, trace=F, control=list(maxit=10000))


summary(ar2ma1garch20)


##
```

```
## Title:
##  GARCH Modelling
##
## Call: garchFit(formula = ~arma(2, 1) + garch(2, 0),
## data = vix, trace = F, control = list(maxit = 10000))
##
## Mean and Variance Equation:
##  data ~ arma(2, 1) + garch(2, 0)
##
## Conditional Distribution:
##  norm
##
## Coefficient(s):
##         mu        ar1        ar2        ma1      omega
##  0.298942   1.000000  -0.018069  -0.150862   0.959320
##    alpha1     alpha2
##  0.424737   0.286003
##
## Std. Errors:
##  based on Hessian
##
## Error Analysis:
##         Estimate  Std. Error  t value Pr(>|t|)
## mu       0.29894         NaN      NaN      NaN
## ar1      1.00000         NaN      NaN      NaN
## ar2     -0.01807         NaN      NaN      NaN
## ma1     -0.15086         NaN      NaN      NaN
## omega    0.95932     0.02592    37.01   <2e-16 ***
## alpha1   0.42474     0.02260    18.80   <2e-16 ***
## alpha2   0.28600     0.01968    14.53   <2e-16 ***
## ---
## Log Likelihood:
##  -14539    normalized:  -1.722
##
## Standardized Residuals Tests:
##                                 Statistic p-Value
##  Jarque-Bera Test   R    Chi^2  15059     0
##  Shapiro-Wilk Test  R    W      NA        NA
##  Ljung-Box Test     R    Q(10)  65.37     3.447e-10
##  Ljung-Box Test     R    Q(15)  70.23     4.058e-09
##  Ljung-Box Test     R    Q(20)  82.93     1.241e-09
##  Ljung-Box Test     R^2  Q(10)  109.8     0
##  Ljung-Box Test     R^2  Q(15)  133.4     0
##  Ljung-Box Test     R^2  Q(20)  168.4     0
##  LM Arch Test       R    TR^2   110.3     0
```

```
##
## Information Criterion Statistics:
##   AIC   BIC   SIC  HQIC
## 3.446 3.452 3.446 3.448
```

```
# Fit ARMA(2,1) + GARCH(3,0)
ar2ma1garch30 <- garchFit(~ arma(2,1) + garch(3, 0),
                    data=vix, trace=F, control=list(maxit=10000))
summary(ar2ma1garch30)
```

```
##
## Title:
##  GARCH Modelling
##
## Call: garchFit(formula = ~arma(2, 1) + garch(3, 0),
## data = vix, trace = F, control = list(maxit = 10000))
##
## Mean and Variance Equation:
##  data ~ arma(2, 1) + garch(3, 0)
##
## Conditional Distribution:
##  norm
##
## Coefficient(s):
##         mu        ar1        ar2        ma1      omega
##  0.208832   1.000000  -0.012859  -0.138628   0.736370
##    alpha1     alpha2     alpha3
##  0.331313   0.230664   0.250680
##
## Std. Errors:
##  based on Hessian
##
## Error Analysis:
##         Estimate  Std. Error  t value Pr(>|t|)
## mu       0.20883     0.06678    3.127  0.00177 **
## ar1      1.00000     0.21206    4.716 2.41e-06 ***
## ar2     -0.01286     0.20850   -0.062  0.95082
## ma1     -0.13863     0.21264   -0.652  0.51444
## omega    0.73637     0.02535   29.046  < 2e-16 ***
## alpha1   0.33131     0.02027   16.342  < 2e-16 ***
## alpha2   0.23066     0.01797   12.838  < 2e-16 ***
## alpha3   0.25068     0.01965   12.760  < 2e-16 ***
## ---
## Log Likelihood:
```

```
##  -14327    normalized:  -1.697
##
## Standardized Residuals Tests:
##                                 Statistic p-Value
##  Jarque-Bera Test   R    Chi^2  12080     0
##  Shapiro-Wilk Test  R    W      NA        NA
##  Ljung-Box Test     R    Q(10)  60.86     2.492e-09
##  Ljung-Box Test     R    Q(15)  68.62     7.847e-09
##  Ljung-Box Test     R    Q(20)  85.86     3.887e-10
##  Ljung-Box Test     R^2  Q(10)  50.23     2.422e-07
##  Ljung-Box Test     R^2  Q(15)  58.53     4.498e-07
##  Ljung-Box Test     R^2  Q(20)  71.4      1.074e-07
##  LM Arch Test       R    TR^2   52.73     4.609e-07
##
## Information Criterion Statistics:
##   AIC   BIC   SIC  HQIC
## 3.396 3.403 3.396 3.399
```

```
# Fit ARMA(2,1) + GARCH(1,1)
ar2ma1garch11 <- garchFit(~ arma(2,1) + garch(1, 1),
                   data=vix, trace=F, control=list(maxit=10000))
```

```
summary(ar2ma1garch11)
```

```
##
## Title:
##  GARCH Modelling
##
## Call: garchFit(formula = ~arma(2, 1) + garch(1, 1),
##  data = vix, trace = F, control = list(maxit = 10000))
##
## Mean and Variance Equation:
##  data ~ arma(2, 1) + garch(1, 1)
##
## Conditional Distribution:
##  norm
##
## Coefficient(s):
##        mu        ar1        ar2        ma1      omega
##  0.247022   1.000000  -0.015715  -0.162233   0.109346
##    alpha1      beta1
##  0.233467   0.745778
##
## Std. Errors:
```

```
##  based on Hessian
##
## Error Analysis:
##         Estimate  Std. Error  t value Pr(>|t|)
## mu      0.247022         NaN      NaN      NaN
## ar1     1.000000         NaN      NaN      NaN
## ar2    -0.015715         NaN      NaN      NaN
## ma1    -0.162233         NaN      NaN      NaN
## omega   0.109346         NaN      NaN      NaN
## alpha1  0.233467    0.012066    19.35   <2e-16 ***
## beta1   0.745778    0.004075   183.00   <2e-16 ***
## ---
## Log Likelihood:
##  -14112    normalized:  -1.672
##
## Standardized Residuals Tests:
##                                 Statistic p-Value
##  Jarque-Bera Test   R    Chi^2  14959     0
##  Shapiro-Wilk Test  R    W      NA        NA
##  Ljung-Box Test     R    Q(10)  64.36     5.377e-10
##  Ljung-Box Test     R    Q(15)  71.26     2.665e-09
##  Ljung-Box Test     R    Q(20)  90.53     5.992e-11
##  Ljung-Box Test     R^2  Q(10)  9.06      0.5264
##  Ljung-Box Test     R^2  Q(15)  12.06     0.6743
##  Ljung-Box Test     R^2  Q(20)  16.95     0.656
##  LM Arch Test       R    TR^2   11.76     0.4649
##
## Information Criterion Statistics:
##   AIC   BIC   SIC  HQIC
## 3.345 3.351 3.345 3.347
```

```
# Fit ARMA(2,1) + GARCH(2,1)
ar2ma1garch21 <- garchFit(~ arma(2,1) + garch(2, 1),
                   data=vix, trace=F, control=list(maxit=10000))
```

```
summary(ar2ma1garch21)
```

```
##
## Title:
##  GARCH Modelling
##
## Call: garchFit(formula = ~arma(2, 1) + garch(2, 1),
##  data = vix, trace = F, control = list(maxit = 10000))
##
```

```
## Mean and Variance Equation:
##  data ~ arma(2, 1) + garch(2, 1)
##
## Conditional Distribution:
##  norm
##
## Coefficient(s):
##          mu          ar1          ar2          ma1
##  0.24705601   0.99999999  -0.01571797  -0.16222267
##       omega       alpha1       alpha2        beta1
##  0.10924265   0.23327699   0.00000001   0.74594883
##
## Std. Errors:
##  based on Hessian
##
## Error Analysis:
##          Estimate  Std. Error  t value Pr(>|t|)
## mu      2.471e-01         NaN      NaN      NaN
## ar1     1.000e+00         NaN      NaN      NaN
## ar2    -1.572e-02         NaN      NaN      NaN
## ma1    -1.622e-01         NaN      NaN      NaN
## omega   1.092e-01         NaN      NaN      NaN
## alpha1  2.333e-01   1.575e-02    14.81   <2e-16 ***
## alpha2  1.000e-08   1.851e-02     0.00        1
## beta1   7.459e-01   6.308e-03   118.25   <2e-16 ***
## ---
## Log Likelihood:
##  -14112    normalized:  -1.672
##
## Standardized Residuals Tests:
##                                  Statistic p-Value
##  Jarque-Bera Test   R    Chi^2  14973     0
##  Shapiro-Wilk Test  R    W      NA        NA
##  Ljung-Box Test     R    Q(10)  64.42     5.225e-10
##  Ljung-Box Test     R    Q(15)  71.32     2.598e-09
##  Ljung-Box Test     R    Q(20)  90.6      5.812e-11
##  Ljung-Box Test     R^2  Q(10)  9.068     0.5257
##  Ljung-Box Test     R^2  Q(15)  12.07     0.6736
##  Ljung-Box Test     R^2  Q(20)  16.97     0.6552
##  LM Arch Test       R    TR^2   11.77     0.4642
##
## Information Criterion Statistics:
##   AIC   BIC   SIC  HQIC
## 3.346 3.352 3.346 3.348
```

```
# Fit ARMA(2,1) + GARCH(3,1)
ar2ma1garch31 <- garchFit(~ arma(2,1) + garch(3, 1),
                    data=vix, trace=F, control=list(maxit=10000))
```

```
summary(ar2ma1garch31)
```

```
##
## Title:
##  GARCH Modelling
##
## Call: garchFit(formula = ~arma(2, 1) + garch(3, 1),
## data = vix, trace = F, control = list(maxit = 10000))
##
## Mean and Variance Equation:
##  data ~ arma(2, 1) + garch(3, 1)
##
## Conditional Distribution:
##  norm
##
## Coefficient(s):
##         mu         ar1         ar2         ma1       omega
## 0.33420432  0.64073786  0.33827662  0.31890709  0.12831399
##     alpha1      alpha2      alpha3       beta1
## 0.22978426  0.00000001  0.00893779  0.72739098
##
## Std. Errors:
##  based on Hessian
##
## Error Analysis:
##         Estimate  Std. Error  t value Pr(>|t|)
## mu     3.342e-01         NaN      NaN      NaN
## ar1    6.407e-01         NaN      NaN      NaN
## ar2    3.383e-01         NaN      NaN      NaN
## ma1    3.189e-01         NaN      NaN      NaN
## omega  1.283e-01   4.671e-03   27.469   <2e-16 ***
## alpha1 2.298e-01   1.548e-02   14.847   <2e-16 ***
## alpha2 1.000e-08   2.082e-02    0.000    1.000
## alpha3 8.938e-03   1.742e-02    0.513    0.608
## beta1  7.274e-01   1.279e-02   56.851   <2e-16 ***
## ---
## Log Likelihood:
##  -14156    normalized:  -1.677
##
```

```
## Standardized Residuals Tests:
##                                  Statistic p-Value
##  Jarque-Bera Test   R    Chi^2  13719     0
##  Shapiro-Wilk Test  R    W      NA        NA
##  Ljung-Box Test     R    Q(10)  83.24     1.161e-13
##  Ljung-Box Test     R    Q(15)  92.05     4.113e-13
##  Ljung-Box Test     R    Q(20)  111.9     8.993e-15
##  Ljung-Box Test     R^2  Q(10)  9.005     0.5317
##  Ljung-Box Test     R^2  Q(15)  11.59     0.7101
##  Ljung-Box Test     R^2  Q(20)  17.43     0.6251
##  LM Arch Test       R    TR^2   11.44     0.4919
##
## Information Criterion Statistics:
##   AIC   BIC   SIC  HQIC
## 3.356 3.364 3.356 3.359
```

```
# Fit ARMA(2,1) + GARCH(1,2)
ar2ma1garch12 <- garchFit(~ arma(2,1) + garch(1, 2),
                    data=vix, trace=F, control=list(maxit=10000))


summary(ar2ma1garch12)
```

```
##
## Title:
##  GARCH Modelling
##
## Call: garchFit(formula = ~arma(2, 1) + garch(1, 2),
## data = vix, trace = F, control = list(maxit = 10000))
##
## Mean and Variance Equation:
##  data ~ arma(2, 1) + garch(1, 2)
##
## Conditional Distribution:
##  norm
##
## Coefficient(s):
##         mu         ar1         ar2         ma1       omega
##  0.1206061   0.9997366  -0.0080361  -0.1402033   0.0871205
##     alpha1       beta1       beta2
##  0.1606271   0.7800181   0.0250442
##
## Std. Errors:
##  based on Hessian
##
```

```
## Error Analysis:
##         Estimate  Std. Error  t value Pr(>|t|)
## mu      0.120606   0.032453    3.716 0.000202 ***
## ar1     0.999737        NaN      NaN      NaN
## ar2    -0.008036        NaN      NaN      NaN
## ma1    -0.140203        NaN      NaN      NaN
## omega   0.087121        NaN      NaN      NaN
## alpha1  0.160627   0.008688   18.488  < 2e-16 ***
## beta1   0.780018   0.079458    9.817  < 2e-16 ***
## beta2   0.025044   0.074130    0.338 0.735482
## ---
## Log Likelihood:
##  -14143    normalized:  -1.675
##
## Standardized Residuals Tests:
##                                 Statistic p-Value
##  Jarque-Bera Test   R    Chi^2  18136     0
##  Shapiro-Wilk Test  R    W      NA        NA
##  Ljung-Box Test     R    Q(10)  69.33     5.981e-11
##  Ljung-Box Test     R    Q(15)  76.01     3.713e-10
##  Ljung-Box Test     R    Q(20)  93.45     1.839e-11
##  Ljung-Box Test     R^2  Q(10)  13.77     0.1837
##  Ljung-Box Test     R^2  Q(15)  17.15     0.31
##  Ljung-Box Test     R^2  Q(20)  20.7      0.4152
##  LM Arch Test       R    TR^2   16.04     0.1892
##
## Information Criterion Statistics:
##   AIC   BIC   SIC  HQIC
## 3.353 3.360 3.353 3.355
```

```
# Fit ARMA(2,1) + GARCH(2,2)
ar2ma1garch22 <- garchFit(~ arma(2,1) + garch(2, 2),
                   data=vix, trace=F, control=list(maxit=10000))
```

```
summary(ar2ma1garch22)
```

```
##
## Title:
##  GARCH Modelling
##
## Call: garchFit(formula = ~arma(2, 1) + garch(2, 2),
## data = vix, trace = F, control = list(maxit = 10000))
##
## Mean and Variance Equation:
```

```
##  data ~ arma(2, 1) + garch(2, 2)
##
## Conditional Distribution:
##  norm
##
## Coefficient(s):
##          mu          ar1          ar2          ma1        omega
##  0.25032680   0.99999999  -0.01591579  -0.16421685   0.12046127
##      alpha1       alpha2        beta1        beta2
##  0.26011114   0.00000001   0.52658677   0.18902572
##
## Std. Errors:
##  based on Hessian
##
## Error Analysis:
##          Estimate  Std. Error  t value Pr(>|t|)
## mu      2.503e-01         NaN      NaN      NaN
## ar1     1.000e+00         NaN      NaN      NaN
## ar2    -1.592e-02         NaN      NaN      NaN
## ma1    -1.642e-01         NaN      NaN      NaN
## omega   1.205e-01   4.387e-03   27.458  < 2e-16 ***
## alpha1  2.601e-01   1.721e-02   15.111  < 2e-16 ***
## alpha2  1.000e-08   2.750e-02    0.000   1.0000
## beta1   5.266e-01   1.019e-01    5.165  2.4e-07 ***
## beta2   1.890e-01   8.608e-02    2.196   0.0281 *
## ---
## Log Likelihood:
##  -14108    normalized:  -1.671
##
## Standardized Residuals Tests:
##                                 Statistic p-Value
##  Jarque-Bera Test   R    Chi^2  14889     0
##  Shapiro-Wilk Test  R    W      NA        NA
##  Ljung-Box Test     R    Q(10)  63.85     6.727e-10
##  Ljung-Box Test     R    Q(15)  70.5      3.64e-09
##  Ljung-Box Test     R    Q(20)  89.64     8.565e-11
##  Ljung-Box Test     R^2  Q(10)  6.903     0.7346
##  Ljung-Box Test     R^2  Q(15)  10.26     0.8033
##  Ljung-Box Test     R^2  Q(20)  15.21     0.7643
##  LM Arch Test       R    TR^2   9.724     0.6401
##
## Information Criterion Statistics:
##   AIC   BIC   SIC  HQIC
## 3.345 3.352 3.345 3.347
```

```
# Fit ARMA(2,1) + GARCH(3,2)
ar2ma1garch32 <- garchFit(~ arma(2,1) + garch(3, 2),
                    data=vix, trace=F, control=list(maxit=10000))


summary(ar2ma1garch32)

##
## Title:
##  GARCH Modelling
##
## Call: garchFit(formula = ~arma(2, 1) + garch(3, 2),
## data = vix, trace = F, control = list(maxit = 10000))
##
## Mean and Variance Equation:
##  data ~ arma(2, 1) + garch(3, 2)
##
## Conditional Distribution:
##  norm
##
## Coefficient(s):
##         mu         ar1         ar2         ma1       omega
## 0.33330001  0.81902867  0.16007835  0.06128491  0.13071756
##     alpha1      alpha2      alpha3       beta1       beta2
## 0.27755252  0.00000001  0.00209541  0.57359224  0.12568376
##
## Std. Errors:
##  based on Hessian
##
## Error Analysis:
##         Estimate  Std. Error  t value Pr(>|t|)
## mu     3.333e-01         NaN      NaN      NaN
## ar1    8.190e-01         NaN      NaN      NaN
## ar2    1.601e-01         NaN      NaN      NaN
## ma1    6.128e-02         NaN      NaN      NaN
## omega  1.307e-01   5.537e-03   23.606  < 2e-16 ***
## alpha1 2.776e-01   1.870e-02   14.840  < 2e-16 ***
## alpha2 1.000e-08   3.246e-02    0.000    1.000
## alpha3 2.095e-03   2.154e-02    0.097    0.923
## beta1  5.736e-01   1.067e-01    5.377 7.55e-08 ***
## beta2  1.257e-01   9.227e-02    1.362    0.173
## ---
## Log Likelihood:
##  -14122    normalized:  -1.673
```

```
##
## Standardized Residuals Tests:
##                                 Statistic p-Value
##  Jarque-Bera Test   R    Chi^2  14156     0
##  Shapiro-Wilk Test  R    W      NA        NA
##  Ljung-Box Test     R    Q(10)  60.73     2.632e-09
##  Ljung-Box Test     R    Q(15)  68.36     8.739e-09
##  Ljung-Box Test     R    Q(20)  88.9      1.154e-10
##  Ljung-Box Test     R^2  Q(10)  7.231     0.7034
##  Ljung-Box Test     R^2  Q(15)  10.15     0.81
##  Ljung-Box Test     R^2  Q(20)  15.59     0.7416
##  LM Arch Test       R    TR^2   10.01     0.6149
##
## Information Criterion Statistics:
##   AIC   BIC   SIC  HQIC
## 3.348 3.357 3.348 3.351
```

```
# Fit ARMA(2,1) + GARCH(1,2)
ar2ma1garch13 <- garchFit(~ arma(2,1) + garch(1, 3),
                  data=vix, trace=F, control=list(maxit=10000))


summary(ar2ma1garch13)
```

```
##
## Title:
##  GARCH Modelling
##
## Call: garchFit(formula = ~arma(2, 1) + garch(1, 3),
## data = vix, trace = F, control = list(maxit = 10000))
##
## Mean and Variance Equation:
##  data ~ arma(2, 1) + garch(1, 3)
##
## Conditional Distribution:
##  norm
##
## Coefficient(s):
##        mu       ar1       ar2        ma1     omega    alpha1
##  0.294648  0.902161  0.079390  -0.033456  0.163279  0.274791
##     beta1     beta2     beta3
##  0.382229  0.241080  0.045156
##
## Std. Errors:
##  based on Hessian
```

```
##
## Error Analysis:
##          Estimate  Std. Error  t value Pr(>|t|)
## mu       0.294648         NaN      NaN      NaN
## ar1      0.902161         NaN      NaN      NaN
## ar2      0.079390         NaN      NaN      NaN
## ma1     -0.033456         NaN      NaN      NaN
## omega    0.163279    0.008203   19.904  < 2e-16 ***
## alpha1   0.274791    0.013932   19.724  < 2e-16 ***
## beta1    0.382229    0.045423    8.415  < 2e-16 ***
## beta2    0.241080    0.080853    2.982  0.00287 **
## beta3    0.045156    0.062018    0.728  0.46655
## ---
## Log Likelihood:
##  -14123    normalized:  -1.673
##
## Standardized Residuals Tests:
##                                 Statistic p-Value
##  Jarque-Bera Test   R    Chi^2  13986     0
##  Shapiro-Wilk Test  R    W      NA        NA
##  Ljung-Box Test     R    Q(10)  59.43     4.654e-09
##  Ljung-Box Test     R    Q(15)  66.18     2.121e-08
##  Ljung-Box Test     R    Q(20)  86.06     3.598e-10
##  Ljung-Box Test     R^2  Q(10)  4.969     0.8932
##  Ljung-Box Test     R^2  Q(15)  7.557     0.9403
##  Ljung-Box Test     R^2  Q(20)  14.49     0.8046
##  LM Arch Test       R    TR^2   7.057     0.8538
##
## Information Criterion Statistics:
##   AIC   BIC   SIC  HQIC
## 3.348 3.356 3.348 3.351
```

```
# Fit ARMA(2,1) + GARCH(2,3)
ar2ma1garch23 <- garchFit(~ arma(2,1) + garch(2, 3),
                    data=vix, trace=F, control=list(maxit=10000))
summary(ar2ma1garch23)
```

```
##
## Title:
##  GARCH Modelling
##
## Call: garchFit(formula = ~arma(2, 1) + garch(2, 3),
## data = vix, trace = F, control = list(maxit = 10000))
##
```

```
## Mean and Variance Equation:
##  data ~ arma(2, 1) + garch(2, 3)
##
## Conditional Distribution:
##  norm
##
## Coefficient(s):
##         mu         ar1         ar2         ma1       omega
## 0.40452254  0.60533256  0.36947811  0.30290178  0.12359318
##     alpha1      alpha2       beta1       beta2       beta3
## 0.23382903  0.00578307  0.72885194  0.00018264  0.00200572
##
## Std. Errors:
##  based on Hessian
##
## Error Analysis:
##         Estimate  Std. Error  t value Pr(>|t|)
## mu      0.4045225   0.1299508    3.113  0.00185 **
## ar1     0.6053326   0.3120248    1.940  0.05238 .
## ar2     0.3694781   0.3047038    1.213  0.22529
## ma1     0.3029018   0.3445207    0.879  0.37929
## omega   0.1235932   0.0042153   29.320  < 2e-16 ***
## alpha1  0.2338290   0.0162880   14.356  < 2e-16 ***
## alpha2  0.0057831   0.0223043    0.259  0.79542
## beta1   0.7288519   0.1136106    6.415  1.4e-10 ***
## beta2   0.0001826   0.1607426    0.001  0.99909
## beta3   0.0020057   0.0700752    0.029  0.97717
## ---
## Log Likelihood:
##  -14138    normalized:  -1.675
##
## Standardized Residuals Tests:
##                                 Statistic p-Value
##  Jarque-Bera Test   R    Chi^2  14285     0
##  Shapiro-Wilk Test  R    W      NA        NA
##  Ljung-Box Test     R    Q(10)  67.24     1.506e-10
##  Ljung-Box Test     R    Q(15)  75.19     5.226e-10
##  Ljung-Box Test     R    Q(20)  96.34     5.648e-12
##  Ljung-Box Test     R^2  Q(10)  8.442     0.5857
##  Ljung-Box Test     R^2  Q(15)  11.14     0.7424
##  Ljung-Box Test     R^2  Q(20)  16.75     0.6688
##  LM Arch Test       R    TR^2   10.94     0.5342
##
## Information Criterion Statistics:
##   AIC   BIC   SIC  HQIC
```

```
## 3.352 3.361 3.352 3.355

# Fit ARMA(2,1) + GARCH(3,3)
ar2ma1garch33 <- garchFit(~ arma(2,1) + garch(3, 3),
                    data=vix, trace=F, control=list(maxit=10000))


summary(ar2ma1garch33)

##
## Title:
##  GARCH Modelling
##
## Call: garchFit(formula = ~arma(2, 1) + garch(3, 3),
## data = vix, trace = F, control = list(maxit = 10000))
##
## Mean and Variance Equation:
##  data ~ arma(2, 1) + garch(3, 3)
##
## Conditional Distribution:
##  norm
##
## Coefficient(s):
##       mu       ar1       ar2       ma1     omega    alpha1
## 0.508519  0.371841  0.596689  0.561623  0.235259  0.257932
##   alpha2    alpha3     beta1     beta2     beta3
## 0.120522  0.082563  0.099013  0.179678  0.208362
##
## Std. Errors:
##  based on Hessian
##
## Error Analysis:
##         Estimate  Std. Error  t value Pr(>|t|)
## mu       0.50852         NaN      NaN      NaN
## ar1      0.37184         NaN      NaN      NaN
## ar2      0.59669         NaN      NaN      NaN
## ma1      0.56162         NaN      NaN      NaN
## omega    0.23526     0.02225    10.57   <2e-16 ***
## alpha1   0.25793     0.01730    14.90   <2e-16 ***
## alpha2   0.12052         NaN      NaN      NaN
## alpha3   0.08256         NaN      NaN      NaN
## beta1    0.09901         NaN      NaN      NaN
## beta2    0.17968         NaN      NaN      NaN
## beta3    0.20836         NaN      NaN      NaN
## ---
```

```
## Log Likelihood:
##  -14149    normalized:  -1.676
##
## Standardized Residuals Tests:
##                                 Statistic p-Value
##  Jarque-Bera Test   R    Chi^2  13730     0
##  Shapiro-Wilk Test  R    W      NA        NA
##  Ljung-Box Test     R    Q(10)  74.81     5.172e-12
##  Ljung-Box Test     R    Q(15)  83.44     1.634e-11
##  Ljung-Box Test     R    Q(20)  105.6     1.235e-13
##  Ljung-Box Test     R^2  Q(10)  7.567     0.671
##  Ljung-Box Test     R^2  Q(15)  10.27     0.8026
##  Ljung-Box Test     R^2  Q(20)  16.52     0.6839
##  LM Arch Test       R    TR^2   10.09     0.6079
##
## Information Criterion Statistics:
##   AIC   BIC   SIC  HQIC
## 3.355 3.364 3.355 3.358
```

It should be noted that most of the *ARMA+GARCH* models are not converged well due to large number of parameters. We will leave this as an exercise to the interested readers to experiment different optimization algorithms in *garchFit* for better model fitting.

8.3.2 Summary of Model Fitting

For model selection, we can table the fitting indices of *AIC*, *BIC*, *SIC* and *HQIC* as in the following table:

```
# Get all the fitting indices
fit.indices =rbind(
ar2ma1garch10@fit$ics,ar2ma1garch20@fit$ics,ar2ma1garch30@fit$ics,
ar2ma1garch11@fit$ics,ar2ma1garch21@fit$ics,ar2ma1garch31@fit$ics,
ar2ma1garch12@fit$ics,ar2ma1garch22@fit$ics,ar2ma1garch32@fit$ics,
ar2ma1garch13@fit$ics,ar2ma1garch23@fit$ics,ar2ma1garch33@fit$ics)
# Match with the models
rownames(fit.indices) = c("ar2ma1garch10",
  "ar2ma1garch20","ar2ma1garch30",
  "ar2ma1garch11","ar2ma1garch21","ar2ma1garch31",
  "ar2ma1garch12","ar2ma1garch22","ar2ma1garch32",
  "ar2ma1garch13","ar2ma1garch23","ar2ma1garch33")
# Print the indices
fit.indices
```

```
##                 AIC   BIC   SIC  HQIC
## ar2ma1garch10 3.540 3.545 3.540 3.542
```

TABLE 8.1
Summary of the 12 ARMA+GARCH Model Fittings

	AIC	BIC	SIC	HQIC
ar2ma1garch10	3.540	3.545	3.540	3.542
ar2ma1garch20	3.446	3.452	3.446	3.448
ar2ma1garch30	3.397	3.403	3.397	3.399
ar2ma1garch11	3.345	3.351	3.345	3.347
ar2ma1garch21	3.346	3.352	3.346	3.348
ar2ma1garch31	3.356	3.364	3.356	3.359
ar2ma1garch12	3.353	3.360	3.353	3.355
ar2ma1garch22	3.345	3.352	3.345	3.347
ar2ma1garch32	3.349	3.357	3.349	3.351
ar2ma1garch13	3.348	3.356	3.348	3.351
ar2ma1garch23	3.352	3.361	3.352	3.355
ar2ma1garch33	3.355	3.364	3.355	3.358

```
## ar2ma1garch20 3.446 3.452 3.446 3.448
## ar2ma1garch30 3.396 3.403 3.396 3.399
## ar2ma1garch11 3.345 3.351 3.345 3.347
## ar2ma1garch21 3.346 3.352 3.346 3.348
## ar2ma1garch31 3.356 3.364 3.356 3.359
## ar2ma1garch12 3.353 3.360 3.353 3.355
## ar2ma1garch22 3.345 3.352 3.345 3.347
## ar2ma1garch32 3.348 3.357 3.348 3.351
## ar2ma1garch13 3.348 3.356 3.348 3.351
## ar2ma1garch23 3.352 3.361 3.352 3.355
## ar2ma1garch33 3.355 3.364 3.355 3.358
```

```
# Load the library of "xtable" to make a table
library(xtable)
# Make the table
knitr::kable(round(xtable(fit.indices),4),
      caption = 'Summary of the 12 ARMA+GARCH Model Fittings',
      booktabs = TRUE )
```

As seen in Table 8.1, the *ARMA(2,1)+GARCH(2,3)* (i.e., *ar2ma1garch23*) have the best fit among all the models which are also converged. The *Ljung-Box Test* for the standardized residuals for the lags of 10, 15, and 20 are all statistically insignificant with *p-values* > 0.5.

This satisfactory model fitting can be visually depicted through the figures below. In Figure 8.4, the left panel displays the Autocorrelation Functions

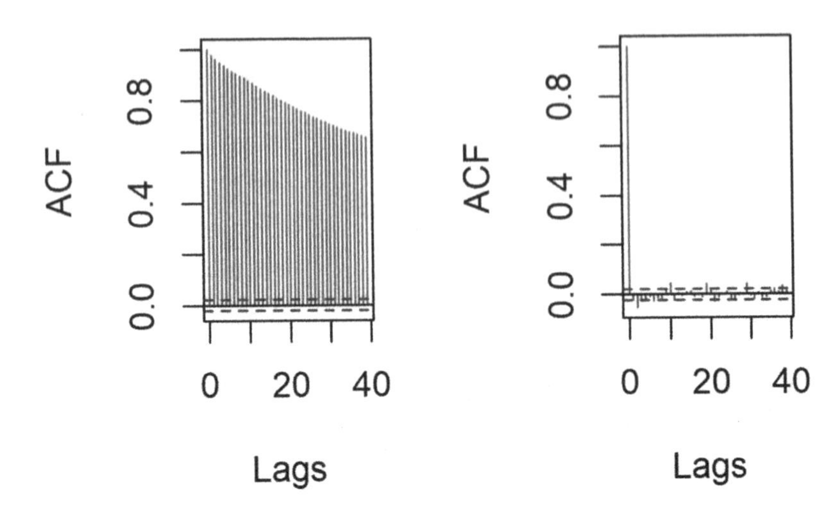

FIGURE 8.4
ACFs for the VIX Data (in the left panel) and the Residuals from ARMA(2,1)+GARCH(2,3) Model (in the right panel)

(ACFs) for the original VIX data, while the right panel exhibits the ACFs for the residuals obtained from the *ARMA(2,1)+GARCH(2,3)* model. In the left panel, all the ACFs for the original VIX data are statistically significant for lags greater than 40. However, the time series autocorrelation concerns have been effectively addressed in the *ARMA(2,1)+GARCH(2,3)* model, as indicated by the absence of statistically significant ACFs in the right panel.

Similarly, Figure 8.5 illustrates the ACFs for the original *squared VIX* data in the left panel and the ACFs for the *squared residuals* from *ARMA(2,1)+GARCH(2,3)* in the right panel. All the ACFs in the left panel for the original squared *VIX* data are statistically significant to the lags of more than 40, but this time series autocorrelation is taken care of in the *ARMA(2,1)+GARCH(2,3)* where all the ACFs are not statistically significant anymore.

In this *ARMA(2,1)+GARCH(2,3)* model, the *Mean* and *Variance* are estimated by the *ARMA(2, 1) + GARCH(2, 3)*, where the parameter estimation can be seen from the following table:

```
# Make the table for parameter estimation
knitr::kable(round(xtable(ar2ma1garch23@fit$matcoef),4),
      caption = 'Parameter Estimation from ARMA(2,1)+GARCH(2,3)',
      booktabs = TRUE
)
```

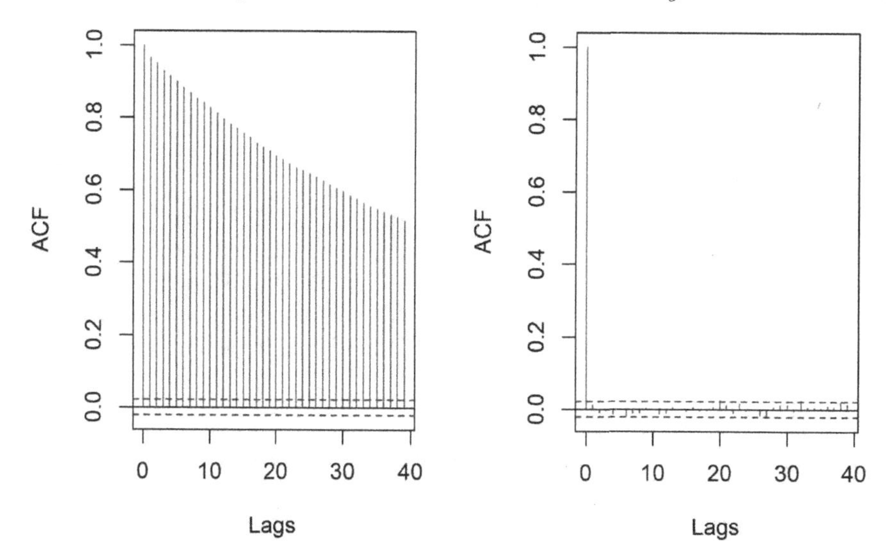

FIGURE 8.5
ACFs for the Squared-VIX Data (in the left panel) and the Squared-Residuals from ARMA(2,1)+GARCH(2,3) Model (in the right panel)

TABLE 8.2
Parameter Estimation from ARMA(2,1)+GARCH(2,3)

	Estimate	Std. Error	t value	Pr(>|t|)
mu	0.4045	0.1300	3.1129	0.0019
ar1	0.6053	0.3120	1.9400	0.0524
ar2	0.3695	0.3047	1.2126	0.2253
ma1	0.3029	0.3445	0.8792	0.3793
omega	0.1236	0.0042	29.3201	0.0000
alpha1	0.2338	0.0163	14.3559	0.0000
alpha2	0.0058	0.0223	0.2593	0.7954
beta1	0.7289	0.1136	6.4154	0.0000
beta2	0.0002	0.1607	0.0011	0.9991
beta3	0.0020	0.0701	0.0286	0.9772

As seen in Table 8.2, some of the parameters associated with *GARCH(2,3)* are not statistically significant, such as α_2, β_2, and β_3. This is an indication that this *GARCH(2,3)* model might be able to simplified to a *GARCH(1,1)* and we leave this as an exercise to our interested readers.

8.4 Discussions and Recommendations

8.4.1 Discussions

This chapter served as an introduction of the *GARCH model* for financial data analysis. To illustrate *GARCH* modeling, we compiled the historical *VIX (Volatility Index)* daily data from January 2nd, 1990 to June 27th, 2023 and analyzed it using the *R* package *fGarch*. We demonstrated the advantages of *GARCH* modeling over the classical *ARIMA* modeling in dealing with volatility dynamics in financial data analysis.

In the data analysis of *VIX* data, we used the default normal error distribution and assumed $\epsilon_t \sim N(0, 1)$ in the *GARCH* modeling throughout the Section 8.2. It has long been noticed that financial data can be *heavy-tailed* or *outlier-prone* due to the non-constant conditional variance or large variance. To handle such non-normal data, we can use other error distributions in the error process to model other error distributions for ϵ_t. This is being taken care of and incorporated in the *R* function *garchFit* in the *fGarch* package. In *garchFit*, we can make change from the default *norm* distribution to other distributions using the option *cond.dist*. In the *garchFit* function, the *cond.dist* argument allows us to specify the *conditional distribution* for the standardized residuals when estimating *GARCH* models. Each distributional assumption corresponds to a different model specification, and the choice can impact the model's fit and interpretation. Here's what each option represents:

- *norm* as the default *Normal Distribution*, which assumes that the conditional innovations (error terms) follow a standard normal (Gaussian) distribution.
- *snorm* as the *Skew Normal Distribution*, which assumes that the conditional innovations follow a skewed normal distribution and is suitable when you suspect the presence of skewness in the distribution of innovations.
- *ged* as the *Generalized Error Distribution (GED)*, which assumes that the conditional innovations follow a generalized error distribution. This distribution provides more flexibility compared to the normal distribution and can capture heavier tails.
- *sged* as the *Skewed Generalized Error Distribution*, which assumes that the conditional innovations follow a skewed generalized error distribution. This distribution combines the features of the *GED* with skewness, allowing for non-Gaussian and asymmetric distributions.
- *std* as the *Student's t-Distribution*, which assumes that the conditional innovations follow a Student's t-distribution. This distribution is useful when you expect fat-tailed or heavy-tailed distributions, which are often observed in financial data.

- *sstd* as the *Skewed Student's t-Distribution*, which assumes that the conditional innovations follow a skewed Student's t-distribution. This distribution is appropriate when you anticipate both skewness and fat tails in the distribution.
- *snig* as the *Skewed Normal Inverse Gaussian Distribution*, which assumes that the conditional innovations follow a skewed normal inverse Gaussian distribution. This distribution is suitable for modeling financial returns data with skewness and leptokurtosis (heavy tails).
- *QMLE* as the *Quasi-Maximum Likelihood Estimation*. This option allows the model to choose the best-fitting distribution automatically based on maximum likelihood estimation. This is a convenient choice when you're unsure about the distribution and want the model to estimate it from the data.

Choosing the appropriate *cond.dist* option depends on the understanding of the data and the characteristics you want to capture in the *GARCH* model. Financial time series data often exhibit characteristics like volatility clustering, fat tails, and skewness, so selecting the right distribution assumption is important for accurate modeling and forecasting. You may need to try multiple distribution assumptions and compare the goodness-of-fit to determine which one works best for your specific dataset.

8.4.2 Recommendations

To conclude this chapter, we provide some recommendations and discussion points below.

8.4.2.1 Why Do We Need GARCH Modeling

The *GARCH* model is a valuable tool in time series analysis for several reasons:

- *Modeling Volatility*: *GARCH* models are specifically designed to model and capture the conditional variance or volatility of a time series. In many real-world applications, financial returns, asset prices, and other economic indicators exhibit changing levels of volatility over time. By using a *GARCH* model, we can better understand and characterize these volatility patterns.
- *Risk Management*: In finance and risk management, accurately modeling and forecasting volatility is crucial for assessing and managing risk. *GARCH* models can help investors, portfolio managers, and financial analysts make more informed decisions by providing insights into the potential range of future price movements.
- *Improved Forecasting*: When modeling financial or economic time series, using a simple *constant variance* assumption as in *ARIMA* models may not be appropriate because it ignores the changing nature of volatility. The

GARCH models provide a more sophisticated framework that can lead to better forecasting accuracy by incorporating past volatility information.

- *Statistical Testing*: *GARCH* models are also useful for conducting statistical tests to determine whether a time series exhibits conditional heteroskedasticity, meaning the variance of the series changes over time based on past observations. Detecting conditional heteroskedasticity is important for understanding the underlying data-generating process and for selecting appropriate models for analysis.
- *Model Diagnostics*: *GARCH* models offer diagnostic tools for assessing the adequacy of model fit. Analysts can examine model residuals to check whether they exhibit autocorrelation or other patterns that indicate inadequacies in the model as what we presented in this chapter. This can guide model refinement and improvement.

In summary, *GARCH* models are essential in time series analysis, particularly in finance and economics, because they help capture and understand the changing nature of volatility in data. They provide a framework for modeling volatility patterns, making better forecasts, managing risk, and improving decision-making in various fields.

8.4.2.2 How to Do ARCH/GARCH Modeling

Choosing the best *GARCH* model fit involves a combination of statistical methods, model evaluation, and domain expertise. Here's a step-by-step guide on how to choose the best *GARCH* model fit for your data:

- *Data Preprocessing*: Start by preparing your financial or time series data. Ensure it is stationary or can be made stationary through differencing, as *GARCH* models are typically applied to stationary series.
- *Model Selection*: Consider different *GARCH* models in *GARCH(pV, qV)* to adjust the numbers pV and qV, which represent the orders of the autoregressive and moving average components of the *GARCH* model. You can use statistical criteria such as the *Akaike Information Criterion (AIC)*, *Bayesian Information Criterion (BIC)*, or *Hannan-Quinn Information Criterion (HQIC)* to compare different *GARCH* models as what we presented in this chapter. Fit multiple *GARCH* models with different combinations of pV and qV and choose the model with the lowest AIC, BIC, or HQIC value. These criteria balance model complexity with goodness of fit. Residual Analysis:
- *Examine the residuals* (model errors) from the *GARCH* model. Residuals should ideally be white noise, indicating that the model captures all relevant information. Check for *autocorrelation* in the residuals using *autocorrelation* and *partial autocorrelation* plots. If autocorrelation is present, it suggests that the model may be missing some important information. Look for significant

patterns or outliers in the residuals, which may indicate model inadequacy or structural changes in the data.

- *Model Diagnostic Tests*: Conduct diagnostic tests, such as the *Ljung-Box* test or the *ARCH-LM* test, to check for the absence of autocorrelation in squared residuals (volatility clustering). Use the *Engle's ARCH test* or the *Ljung-Box test* to assess the adequacy of the *GARCH* model's fit to the squared residuals.
- *Domain Knowledge*: Incorporate domain expertise, if available. Some financial time series may exhibit unique characteristics that are not fully captured by statistical criteria. Expert insights can guide the model selection process.
- *Iterative Refinement*: The process of selecting the best *GARCH* model may involve iterative refinement. Experiment with different model specifications, evaluate their performance, and adjust as needed until you achieve a satisfactory fit.

In summary, choosing the best *GARCH* model fit involves a combination of statistical criteria, model evaluation techniques, and subject matter expertise. It's essential to balance between model complexity and goodness of fit while ensuring that the chosen model effectively captures the volatility patterns in your time series data. For further reading, we recommend the book by Ruppert and Matteson (2015).

8.5 Exercises

1. Analyze the daily *HIGH* VIX data to examine whether there is significant heteroskedasticity for *HIGH* VIX data, then model it with *GARCH* model.
2. Analyze the daily *LOW* VIX data to examine whether there is significant heteroskedasticity for *LOW* VIX data, then model it with *GARCH* model.
3. Analyze the daily *CLOSE* VIX data to examine whether there is significant heteroskedasticity for *CLOSE* VIX data, then model it with *GARCH* model.
4. Examine the correlations among these 4 *VIX* time series. (This will be used for Chapter 9 of *Cointegration*).

9

Cointegration

The typical *AutoRegressive Integrated Moving-Average(ARIMA)* models discussed in Chapter 7 are employed to analyze and predict the behavior of a single univariate time series. They consist of *autoregressive (AR)*, *differencing (I for integrated)*, and *moving-average (MA)* components. *ARIMA* models are useful for capturing and forecasting short- to medium-term trends, seasonality, and fluctuations in a single time series. *ARIMA* models are often used for tasks like stock price forecasting or economic indicator prediction. However, they are limited in their ability to address the relationships between multiple time series or assets.

Furthermore, the *Autoregressive Conditional Heteroskedasticity (ARCH)* and *Generalized Autoregressive Conditional Heteroskedasticity (GARCH)* models discussed in Chapter 8 are focused on modeling the conditional volatility of a single time series. They are particularly relevant for capturing and forecasting the changing volatility patterns and heteroskedasticity within a univariate time series. *GARCH* models are widely used in financial markets for risk assessment, option pricing, and modeling asset returns. Like *ARIMA* models, *GARCH* models also deal with individual time series and do not consider the interactions or long-term relationships between multiple assets or variables.

Transitioning from *ARIMA* and *GARCH* models to *cointegration* in this chapter, we will introduce *cointegration* for analyzing long-term relationships and interactions between multiple time series so that a *stationary* process can be generated from multiple time series that are *non-stationary* as discussed in Chapter 7. For this purpose, we will use two datasets to illustrate how to use *R* packages for *cointegration analysis.*

9.1 What *R* Packages for *Cointegration*

Cointegration analysis can be performed in *R* using several packages. Some of the commonly used R packages for cointegration analysis include:

- *urca (Unit Root and Cointegration Tests)*: The *urca* package is specifically designed for unit root and cointegration testing, which is maintained by

DOI: 10.1201/9781003469704-9

Bernhard Pfaff. It provides functions for the *Engle-Granger* and *Johansen* cointegration tests as detailed in Pfaff (2008). The functions *ca.jo()* is for *Johansen cointegration test* and *ca.po()* for Phillips-Ouliaris cointegration test. More information can be found from its *help* as follows *library(help=urca)*.

- *vars (Vector Autoregression Models)*: The *vars* package allows us to estimate *Vector Error Correction Models (VECM)* when cointegration is detected. It's particularly useful for modeling and analyzing multivariate time series data with function *VECM()* for estimating VECMs. As an extension of the *vars* package, *vars2 (VAR Modelling)* includes additional tools for estimating vector autoregressive models and VECMs. Details can be found from its *help* as follows *library(help=vars)*.
- *tsDyn (Time Series Analysis in Dynamic Linear Models)*: The *tsDyn* package provides tools for time series modeling and includes functions for cointegration analysis, including threshold cointegration. The functions of *VECM()* are for estimating VECMs and *lineVar()* for estimating linear cointegrating models.
- *tseries (Time Series Analysis and Computational Finance)*: The *tseries* package offers a wide range of time series analysis tools, including unit root tests, which are often used as a preliminary step in cointegration analysis, where *adf.test()* can be used for *Augmented Dickey-Fuller tests* and *pp.test()* for Phillips-Perron tests.
- *Ecdat (Data Sets for Econometrics)*: The *Ecdat* package includes several datasets for econometric analysis, some of which can be used for cointegration exercises.

These packages offer various functions and methods for performing cointegration tests and modeling cointegrated systems. Depending on your specific analysis requirements and preferences, you may choose one or more of these packages to conduct cointegration analysis in *R*. We will mainly use the *urca* package in this chapter.

9.2 Statistical Methods

Before we perform *cointegration analysis* for real data, let's briefly describe and discuss the associated statistical methods and statistical tests without getting into too much mathematical detail.

9.2.1 The Augmented Dickey-Fuller Test for Stationarity

The *Augmented Dickey-Fuller (ADF)* test is a statistical test on whether a time series dataset is *stationary* or possesses a *unit root*, which indicates *non-stationarity.* Stationary time series data is essential for many financial time series analysis techniques and forecasting models. The *ADF* test builds upon the simpler *Dickey-Fuller test* by including lagged differences of the dependent variable in the regression equation.

In the *ADF test for unit root*, the *Null Hypothesis* (H_0) is that the time series has a unit root, which means it is non-stationary and the *Alternative Hypothesis* (i.e. H_1) is that the time series is stationary (i.e., it does not have a unit root).

Based on the hypothesis, a test statistic is calculated from *ADF* test to measure how strongly the data series reverts to a mean. The test statistic is compared to critical values from a table to determine whether to reject the *null hypothesis.* The *ADF* test uses an *ADF* regression equation that includes lagged differences of the dependent variable. The equation generally takes the following form:

$$\Delta Y_t = \alpha + \beta t + \gamma Y_{t-1} + \delta_1 \Delta Y_{t-1} + \delta_2 \Delta Y_{t-2} + \cdots + \delta_p \Delta Y_{t-p} + \epsilon_t \quad (9.1)$$

where

- ΔY_t represents the differenced time series,
- α is a constant term,
- βt represents a time trend, if applicable,
- γY_{t-1} captures the impact of the lagged value of the time series,
- $\delta_i \Delta Y_{t-i}$ includes lagged differences of the time series up to a certain lag order (p),
- ϵ_t is the error term.

The *ADF* test involves comparing the calculated test statistic with critical values for various confidence levels at *1pct, 5pct, 10pct.* If the test statistic is more negative than the critical values, we can then reject the *null hypothesis* in favor of *stationarity.* If the *null hypothesis* is rejected, it suggests that the time series is stationary, which means it does not have a unit root. If you fail to reject the null hypothesis, it suggests that the time series is non-stationary and likely has a unit root.

This *ADF test* is a widely used tool in econometrics and time series analysis to assess the stationarity of data and is especially important when dealing with financial and economic time series data for modeling and forecasting purposes.

9.2.2 Johansen Test Procedure

Johansen procedure allows us to analyze whether two or more time series can form a *cointegrating* relationship. In quantitative finance or trading, this would allow us to form a portfolio of two or more securities in a mean reversion trading strategy.

A more theoretical explanation of the Johansen test will require some understanding of multivariate time series, mostly to *Vector Autoregressive Models (VAR)*, which are a multidimensional extension of the *Autoregressive Models* (i.e., *AR(p)*) described in Chapter 7 except that each quantity in *VAR(p)* is now vector valued with multiple time series and the coefficients are now matrices. The general form of the *VAR(p)* model, without drift, is given by:

$$Y_t = \mu + A_1 Y_{t-1} + \cdots + A_p Y_{t-p} + \epsilon_t \tag{9.2}$$

where

- μ is the vector-valued mean of the series,
- A_i is the coefficient matrices for each lag and
- ϵ_t is a multivariate Gaussian noise term with mean zero.

With this *VAR(p)*, we can then form a *Vector Error Correction Model (VECM)* by differencing the series:

$$\Delta Y_t = \mu + A Y_{t-1} + \Gamma_1 \Delta Y_{t-1} + \cdots + \Gamma_p \Delta_p Y_{t-p} + \epsilon_t \tag{9.3}$$

where

- $\Delta Y_t = Y_t - Y_{t-1}$ is the differencing operator,
- A is the coefficient matrix for the first lag and
- Γ_i are the matrices for each differenced lag.

The test for no cointegration is then to test whether the matrix A =0.

The *Johansen test* is then derived to check for multiple linear combinations of time series for forming stationary series. This is done by the so-called matrix eigenvalue decomposition. The rank of the A matrix is given by and the *Johansen test* sequentially tests whether this rank is equal to zero, equal to one, through to $r = N - 1$ (where N is the number of time series under test). Then, if $r = 0$ is not rejected, that means that there is no cointegration at all and a rank $r > 0$ implies a cointegrating relationship between two or possibly more time series.

With the eigenvalue decomposition, we can get a set of eigenvectors. The components of the largest eigenvector signal the important property of forming the coefficients of a linear combination of time series to produce a new combined stationary series.

9.3 Standard & Poor Data

9.3.1 Descriptive Analysis

To illustrate the principle of cointegration, we will use the publicly available dataset *SandPhedge.csv*, which contains monthly returns $Y_t = (y_{1t}, y_{2t})'$, where:

1. y_{1t} = the S&P500 index (in column 2 named *Spot*),
2. y_{2t} = the S&P500 futures (in column 3 named *Futures*), and
3. t = the associated time (in column 1 named *Date*) in *month* from February 2002 to April 2013 with 135 observations for $t = 1, \cdots, 135$.

More details about this data can be found in Tao and Brooks (2019). The data can be publicly downloaded (this data was downloaded on October 1st, 2023) from https://www.cambridge.org/us/universitypress/textbooks/introductory-econometrics/eviews-data.

With the downloaded data in *.csv* format, we can read it into *R* and name it as *dSPH* as follows:

```
# Read in the data
dSPH = read.csv("data/SandPHedge.csv",header=T)
# Check the dimension
dim(dSPH)
```

```
## [1] 135   3
```

```
# Print the first 6 observations
head(dSPH)
```

```
##      Date   Spot Futures
## 1 Feb-02 1106.7  1106.9
## 2 Mar-02 1147.4  1149.2
## 3 Apr-02 1076.9  1077.2
## 4 May-02 1067.1  1067.5
## 5 Jun-02  989.8   990.1
## 6 Jul-02  911.6   911.5
```

Note that the *Date* variable in dataset *dSPH* is a character variable in the format of *month-year* and we need to convert it into *Dates* for time series analysis. For this purpose, we will make use of *R* package *lubridate* to convert the *character* formats into *Dates*. This can be done using the function *my()*

(*my* stands for *month-year*) to create a new variable *date* as in the following *R* code chunk:

```
# Create a dataframe
dSPH = data.frame(dSPH)
# Make a new variable: date
dSPH$date = lubridate::my(dSPH$Date)
# Print the first 6 observations
head(dSPH)
```

```
##      Date   Spot Futures       date
## 1 Feb-02 1106.7  1106.9 2002-02-01
## 2 Mar-02 1147.4  1149.2 2002-03-01
## 3 Apr-02 1076.9  1077.2 2002-04-01
## 4 May-02 1067.1  1067.5 2002-05-01
## 5 Jun-02  989.8   990.1 2002-06-01
## 6 Jul-02  911.6   911.5 2002-07-01
```

With the new data, we can now make some time series plots to check the time series autocorrelation. We will plot the time series below for both *Spot* and *Futures* with their *acf* and *pacf*:

```
# Figure layout
par(mfrow=c(2,3),mar=c(4,4,1,2),oma=c(1,1,1,1))
# Plots for "Spot"
plot(Spot~date,type="b", dSPH)
acf(dSPH$Spot, main="");
pacf(dSPH$Spot, main="")
# Plot for "Futures
plot(Futures~date,type="b", dSPH)
acf(dSPH$Futures, main="");
pacf(dSPH$Futures, main="")
```

In Figure 9.1, the three plots in the first row are for *Spot* and the three plots in the second row are for *Futures*. As seen from Figure 9.1, both time series fluctuate without distinctive *trends* (in column 1). However, there is a significant at least order-1 time series autocorrelation (in columns 2 and 3).

9.3.2 Test for Non-Stationarity

To test whether there is a stationarity for both series, we can use the *Augmented Dickey Fuller Unit Root Tests for Stationary* which is implemented in *R* function *ur.df* as follows:

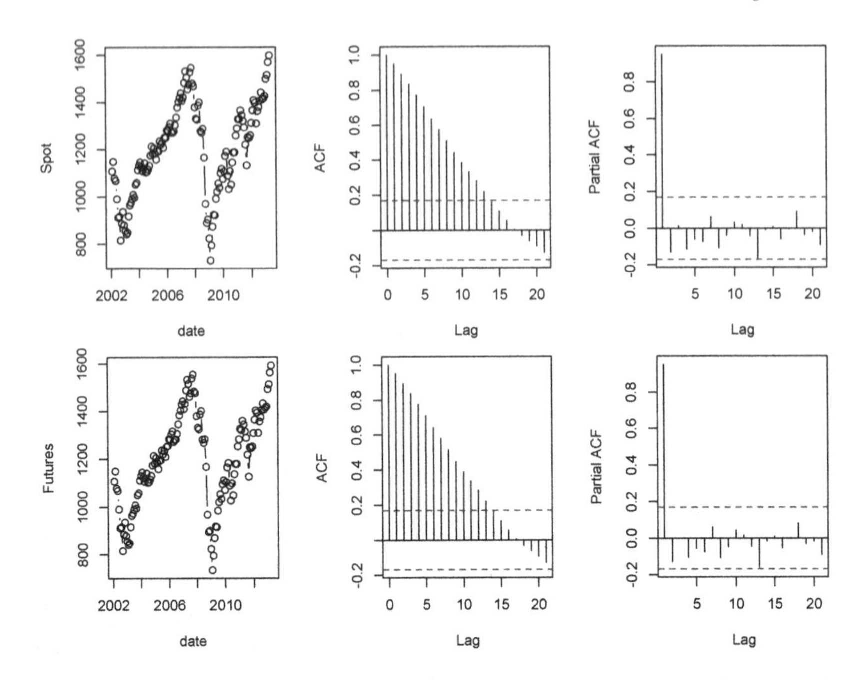

FIGURE 9.1
Time Series Plot along with ACF and PACF

```
# Load the library urca
library(urca)
# Call ur.df to test "Spot" data
test.Spot = ur.df(dSPH$Spot, type = "none",
                  selectlags = "AIC" )
# Print the summary of the test
summary(test.Spot)
```

```
##
## ###############################################
## # Augmented Dickey-Fuller Test Unit Root Test #
## ###############################################
##
## Test regression none
##
##
## Call:
## lm(formula = z.diff ~ z.lag.1 - 1 + z.diff.lag)
##
```

```
## Residuals:
##      Min       1Q  Median      3Q     Max
## -173.29  -25.31    8.35   30.11  135.11
##
## Coefficients:
##             Estimate Std. Error t value Pr(>|t|)
## z.lag.1      0.00153    0.00355    0.43    0.667
## z.diff.lag  0.19401    0.08588    2.26    0.026 *
## ---
## Signif. codes: 0 '***' 0.001 '**' 0.01 '*' 0.05 '.' 0.1 ' ' 1
##
## Residual standard error: 49.4 on 131 degrees of freedom
## Multiple R-squared:  0.0405, Adjusted R-squared:  0.0259
## F-statistic: 2.77 on 2 and 131 DF,  p-value: 0.0666
##
##
## Value of test-statistic is: 0.4315
##
## Critical values for test statistics:
##       1pct  5pct 10pct
## tau1 -2.58 -1.95 -1.62
```

Note that the *R* function *ur.df()* computes the augmented Dickey-Fuller test. We set the *type* to *none* since there is no *trend* or *drift* as seen in Figure 9.1. As seen from the summary of the test, The value of test-statistic (0.4315) is lower than of three critical values for the test statistics (in absolute term) of 2.58(1pct), 1.95(5pct) and 1.62(10pct). So *Spot* is not stationary based on the *Dickey-Fuller test.*

Similarly, we can test for stationarity for the time series *Futures* as follows:

```
# Call ur.df to test "Futures" data
test.Futures = ur.df(dSPH$Futures, type = "none",
                     selectlags = "AIC" )
# Print the summary of the test
summary(test.Futures)
```

```
##
## ###############################################
## # Augmented Dickey-Fuller Test Unit Root Test #
## ###############################################
##
## Test regression none
##
##
## Call:
```

```
## lm(formula = z.diff ~ z.lag.1 - 1 + z.diff.lag)
##
## Residuals:
##      Min       1Q   Median       3Q      Max
## -179.08   -25.46     7.58    29.20   139.74
##
## Coefficients:
##            Estimate Std. Error t value Pr(>|t|)
## z.lag.1     0.00149    0.00353    0.42    0.674
## z.diff.lag  0.19749    0.08581    2.30    0.023 *
## ---
## Signif. codes: 0 '***' 0.001 '**' 0.01 '*' 0.05 '.' 0.1 ' ' 1
##
## Residual standard error: 49 on 131 degrees of freedom
## Multiple R-squared:  0.0418, Adjusted R-squared:  0.0272
## F-statistic: 2.86 on 2 and 131 DF,  p-value: 0.061
##
##
## Value of test-statistic is: 0.4214
##
## Critical values for test statistics:
##       1pct  5pct 10pct
## tau1 -2.58 -1.95 -1.62
```

Again in absolute terms, the value of test statistic (0.4214) is lower than of three critical values of 2.58(1pct), 1.95(5pct) and 1.62(10pct). So the time series *Futures* is also not *stationary*. Therefore both time series are *I(1)*, (i.e., *integrated* of order 1 as discussed in Chapter 7), which leads to the investigation of cointegration.

9.3.3 Cointegration Regression

Cointegration is a statistical concept used to analyze the long-term relationship between two or more time series variables. In essence, it helps determine whether two or more *non-stationary* time series move together in a stable manner over time. When two time series are cointegrated, it means they share a common stochastic pattern. This is often the case when dealing with financial or economic data, where multiple variables (such as stocks) may be related in the long run, even if they exhibit short-term fluctuations.

One common way to test for *cointegration* is to perform a regression between the two time series and then examine the stationarity of the residuals. This can be done using *unit root tests* like the *Dickey-Fuller* test. The null hypothesis of the Dickey-Fuller test is that the time series has a unit root, which implies it is *non-stationary*. If the test rejects the null hypothesis (i.e., the test statistic falls below a critical value), it suggests that the time series is stationary or integrated of order 0 (i.e., *I(0)*), indicating that cointegration may exist between the two series.

Therefore, in a *cointegration test*, we want to reject the null hypothesis of non-stationarity (i.e., the presence of a unit root) in the residuals. This is because, in the context of cointegration, we are trying to demonstrate that the combination of the two non-stationary time series results in a stationary series (i.e., the residuals).

The above discussions can be implemented with the following *R* code chunk:

```
# Run the linear regression
lmSPH = lm(Futures~Spot, dSPH)
# Print the summary of regression
summary(lmSPH)
```

```
##
## Call:
## lm(formula = Futures ~ Spot, data = dSPH)
##
## Residuals:
##     Min      1Q Median     3Q    Max
## -8.732 -2.696  0.453  1.817  9.802
##
## Coefficients:
##             Estimate Std. Error t value Pr(>|t|)
## (Intercept) -4.95072    2.29544   -2.16    0.033 *
## Spot         1.00391    0.00189  530.85   <2e-16 ***
## ---
## Signif. codes: 0 '***' 0.001 '**' 0.01 '*' 0.05 '.' 0.1 ' ' 1
##
## Residual standard error: 4.27 on 133 degrees of freedom
## Multiple R-squared:     1,    Adjusted R-squared:     1
## F-statistic: 2.82e+05 on 1 and 133 DF,  p-value: <2e-16
```

```
# Extract the residuals
res.lmSPH = resid(lmSPH)
# Call Dickey-Fuller stationary test
test.lmSPH = ur.df(res.lmSPH, type = "none",
                   selectlags = "AIC" )
summary(test.lmSPH)
```

```
##
## ###############################################
## # Augmented Dickey-Fuller Test Unit Root Test #
```

```
## ###############################################
##
## Test regression none
##
##
## Call:
## lm(formula = z.diff ~ z.lag.1 - 1 + z.diff.lag)
##
## Residuals:
##    Min     1Q Median     3Q    Max
## -7.822 -2.267 -0.078  1.137  7.673
##
## Coefficients:
##            Estimate Std. Error t value Pr(>|t|)
## z.lag.1     -0.2410     0.0734   -3.28   0.0013 **
## z.diff.lag  -0.3520     0.0830   -4.24  4.2e-05 ***
## ---
## Signif. codes: 0 '***' 0.001 '**' 0.01 '*' 0.05 '.' 0.1 ' ' 1
##
## Residual standard error: 3.19 on 131 degrees of freedom
## Multiple R-squared:  0.286,  Adjusted R-squared:  0.275
## F-statistic: 26.2 on 2 and 131 DF,  p-value: 2.69e-10
##
##
## Value of test-statistic is: -3.283
##
## Critical values for test statistics:
##       1pct  5pct 10pct
## tau1 -2.58 -1.95 -1.62
```

As seen from the output. the test statistic value of 3.280895 (in absolute value) is higher than of 3 critical values of 2.58 (1pct), 1.95(5pct) and 1.62(10pct), indicating that the *null hypothesis* of *non-stationary* is rejected. If residuals are *stationary*, then both series are coinciding and there exists some long-term relationship between variables.

Though stationary, the residuals still have a large amount of autocorrelation and may have a long-term memory, which can be graphically demonstrated in Figure 9.2 produced with the following *R* code chunk:

```
# Figure layout
layout(matrix(c(1,1,2,3),2,2,byrow=T))
dSPH$res = res.lmSPH
# Plots for "Residuals"
plot(res~date, dSPH,type="b",
```

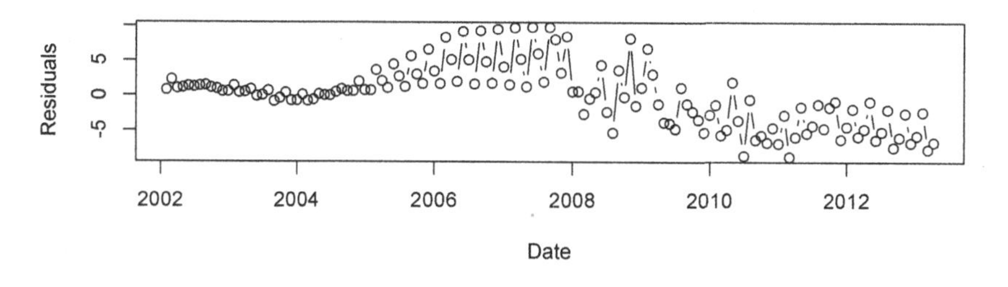

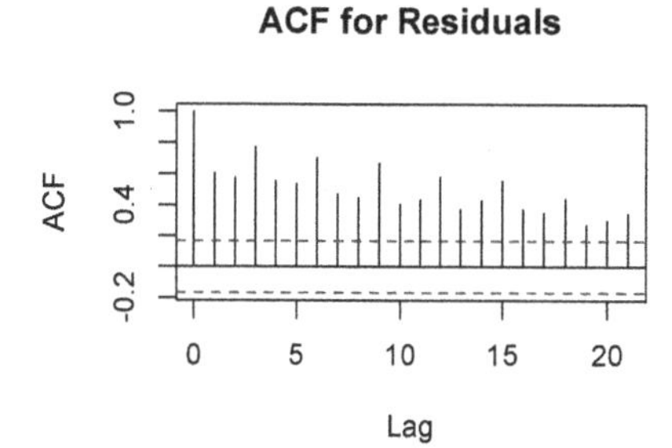

FIGURE 9.2
Time Series Plot for the Residuals

```
     xlab="Date",ylab="Residuals",
     main="Time Series Plot for Regression Residuals")
acf(res.lmSPH, main="ACF for Residuals");
pacf(res.lmSPH, main="PACF for Residuals")
```

9.3.4 Vector Error Correction Model

Another test for *cointegration* is the *Johansen* Procedure for VAR via *ca.jo*. Since we only have two univariate time series of y_{1t} for *Spot* and y_{2t} for *Futures*, we can only have two ranks: $r = 0$ or $r \leq 1$. This means that either there is no cointegration ($r = 0$) or there is ($r \leq 1$).

This can be tested using *Johansen Procedure for VAR* with *R* function *ca.jo* as follows. This function has several options:

- The *type* parameter tells the function whether to use the *trace test statistic* or the *maximum eigenvalue test statistic*, which are the two separate forms of the *Johansen test*. In this example, we will use *trace*.

- K is the number of lags to use in the *VAR* model and is set to the minimum, $K=2$.
- *ecdet* refers to whether to use a *constant* or *drift* term in the model,
- *spec="longrun"* refers to the specification of the *VECM* discussed above. This parameter can be specified as *spec="longrun"* or *spec="transitory"*.

```
# Load the "urca" library
library(urca)
# Perform ca.jo test
cajo.test = ca.jo(dSPH[,c(2,3)], type="trace",
                  spec="longrun")
# Print the summary of the test
summary(cajo.test)
```

```
##
## ######################
## # Johansen-Procedure #
## ######################
##
## Test type: trace statistic , with linear trend
##
## Eigenvalues (lambda):
## [1] 0.08374 0.01137
##
## Values of teststatistic and critical values of test:
##
##           test 10pct  5pct  1pct
## r <= 1 |  1.52  6.50  8.18 11.65
## r = 0  | 13.15 15.66 17.95 23.52
##
## Eigenvectors, normalized to first column:
## (These are the cointegration relations)
##
##              Spot.l2 Futures.l2
## Spot.l2       1.0000        1.0
## Futures.l2 -0.9945      -30.8
##
## Weights W:
## (This is the loading matrix)
##
##             Spot.l2 Futures.l2
## Spot.d       0.1352  0.0009191
## Futures.d    0.3839  0.0009059
```

As seen from the output, we have *Eigenvalues (lambda)* as *0.08374389* and *0.01137206* as the eigenvalues generated by the test with the largest eigenvalue to be *0.08374389.*

The next part is the *Values of test statistic and critical values of test* as below:

```
          test 10pct  5pct  1pct
r <= 1 |  1.52  6.50  8.18 11.65
r = 0  | 13.15 15.66 17.95 23.52
```

This is to test the null hypotheses of $r = 0$ and $r \leq 1$ where the column *test* gives the value of the statistic associated with the critical values at the confidence of 10%, 5%, and 1% respectively. Since the value of test statistic of 13.15 for $r = 0$ (i.e., for null hypothesis of *nointegration*) with the critical values *15.66, 17.95, 23.52*, we can conclude that this null hypothesis is marginally not rejected statistically. Similarly, the null hypothesis of $r \leq 1$ is not rejected statistically. This seems contradictory with the *Augmented Dickey-Fuller Test Unit Root Test* carried out by *R* function *ur.df* in Section 9.3.3. However to notice that in the *Johansen test*, the linear combination values are estimated as part of the test, which can have less statistical power to identify the *cointegration* than the *Augmented Dickey-Fuller Test Unit Root Test* especially for small sample size (we only have 135 observations in this data), and also the *Johsnsen test* here is marginally significant for *cointegration.*

Therefore, let's proceed with forming a *linear combination* of two non-stationary time series to form a new combined *stationary series.* This can be done by using the information from the component:

```
 Eigenvectors, normalized to first column:
(These are the cointegration relations)


               Spot.12 Futures.12
Spot.12      1.0000000    1.00000
Futures.12 -0.9944617  -30.80185
```

We can use the first *eigenvector* column associated with the largest eigenvalue *0.08374389*, which is *(1, -0.9944617).* Therefore, the new linear combination:

$$Z_t = y_{1t} - 0.994 \times y_{2t} = \text{Spot}_t - 0.994 \times \text{Futures}_t \tag{9.4}$$

will be a stationary series.

To confirm this resultant marginal stationary series of Z_t, we can call *R* function *ur.df* for an *Augmented Dickey-Fuller Test Unit Root Test* of stationary. This can be done as follows:

```
# Form the new linear combination series
zt = dSPH$Spot - 0.9944617 *dSPH$Futures
```

```
# Call the Dickey-Fuller Test
test.linComb = ur.df(zt, type = "none",
                     selectlags = "AIC" )
# Print the test summary
summary(test.linComb)
```

```
##
## ###############################################
## # Augmented Dickey-Fuller Test Unit Root Test #
## ###############################################
##
## Test regression none
##
##
## Call:
## lm(formula = z.diff ~ z.lag.1 - 1 + z.diff.lag)
##
## Residuals:
##    Min     1Q Median     3Q    Max
## -7.211 -0.934  0.672  2.139  7.979
##
## Coefficients:
##            Estimate Std. Error t value Pr(>|t|)
## z.lag.1     -0.0440     0.0368   -1.20     0.23
## z.diff.lag  -0.4487     0.0792   -5.67  8.9e-08 ***
## ---
## Signif. codes: 0 '***' 0.001 '**' 0.01 '*' 0.05 '.' 0.1 ' ' 1
##
## Residual standard error: 3.3 on 131 degrees of freedom
## Multiple R-squared:  0.232,  Adjusted R-squared:  0.221
## F-statistic: 19.8 on 2 and 131 DF,  p-value: 3.04e-08
##
##
## Value of test-statistic is: -1.196
##
## Critical values for test statistics:
##       1pct  5pct 10pct
## tau1 -2.58 -1.95 -1.62
```

As seen from the output, the test statistic value of 1.1955 (in absolute value) is higher than the critical value of 1.62 (10pct), but not the critical values of 2.58 (1pct) and 1.95 (5pct), again with a marginal significant at 10% confidence level for stationary cointegration. This property can be graphically illustrated

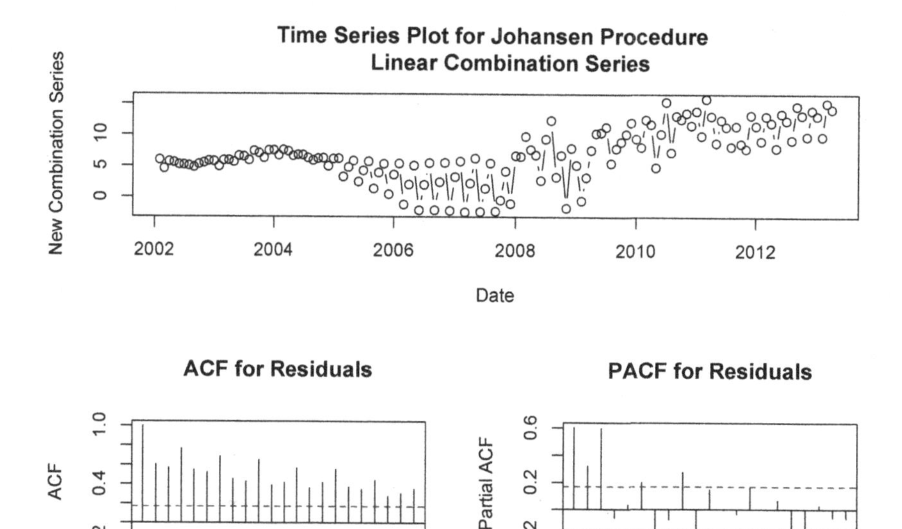

FIGURE 9.3
Time Series Plot for the Linear Combination from Johansen Procedure for VAR

in Figure 9.3 with following *R* code chunk, which behaves similarly to the Figure 9.2.

```
# Figure layout
layout(matrix(c(1,1,2,3),2,2,byrow=T))
# Add zt to the dataframe dSPH
dSPH$zt = zt
# Plots for "Residuals"
plot(zt~date, dSPH,type="b",
     xlab="Date",ylab="New Combination Series",
     main="Time Series Plot for Johansen Procedure
     Linear Combination Series")
acf(zt, main="ACF for Residuals");
pacf(zt, main="PACF for Residuals")
```

9.4 The Volatility Index Data

We re-analyze the *VIX* data from Chapter 8, which is a daily data with 8,441 observations of the daily *OPEN*, *HIGH*, *LOW*, and *CLOSE* indexes with $Y_t = (y_{1t}, y_{2t}, y_{3t}, y_{4t})'$ as:

- y_{1t} is the VIX *OPEN* index at time t,
- y_{2t} is the VIX *HIGH* index at time t,
- y_{3t} is the VIX *LOW* index at time t,
- y_{4t} is the VIX *CLOSE* index at time t,
- t is the associated time from t =1 (*1/2/1990*) to t = 8,441 (*6/27/2023*).

We can load the data into *R* and create a new variable *Date* as what we did in Chapter 8:

```
# Read the data into R
dVIX = read.csv("data/VIX_history.csv", header=T)
# Create the Date variable
dVIX$Date = as.Date(dVIX$DATE, format = "%m/%d/%Y")
# Print the dimension of the data
dim(dVIX)
```

```
## [1] 8441    6
```

```
# Print the first 6 observations
head(dVIX)
```

```
##        DATE  OPEN  HIGH   LOW CLOSE       Date
## 1 1/2/1990 17.24 17.24 17.24 17.24 1990-01-02
## 2 1/3/1990 18.19 18.19 18.19 18.19 1990-01-03
## 3 1/4/1990 19.22 19.22 19.22 19.22 1990-01-04
## 4 1/5/1990 20.11 20.11 20.11 20.11 1990-01-05
## 5 1/8/1990 20.26 20.26 20.26 20.26 1990-01-08
## 6 1/9/1990 22.20 22.20 22.20 22.20 1990-01-09
```

9.4.1 Descriptive Analysis

With the *dVIX* data, we can now make some time series plots to check the time series trends and their autocorrelations as seen in the follow figure:

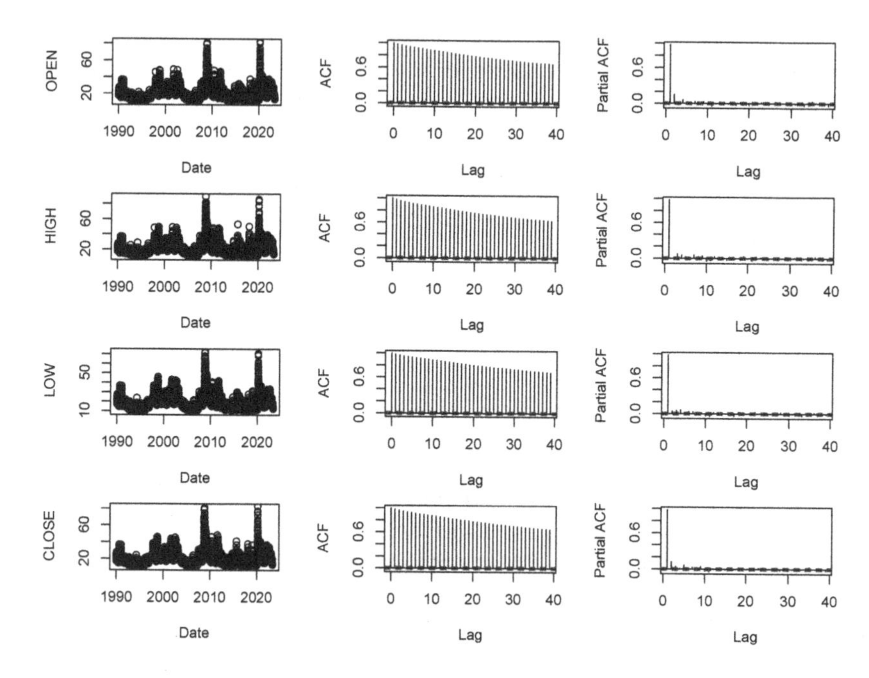

FIGURE 9.4
VIX Time Series Plot along with ACF and PACF

```
# Figure layout
par(mfrow=c(4,3),mar=c(4,4,1,2),oma=c(1,1,1,1))
# Plots for "OPEN"
plot(OPEN~Date,type="b", dVIX)
acf(dVIX$OPEN, main="");pacf(dVIX$OPEN, main="")
# Plots for "HIGH"
plot(HIGH~Date,type="b", dVIX)
acf(dVIX$HIGH, main="");pacf(dVIX$HIGH, main="")
# Plots for "LOW"
plot(LOW~Date,type="b", dVIX)
acf(dVIX$LOW, main="");pacf(dVIX$LOW, main="")
# Plots for "CLOSE"
plot(CLOSE~Date,type="b", dVIX)
acf(dVIX$CLOSE, main="");pacf(dVIX$CLOSE, main="")
```

In Figure 9.4, we can see that all four time series fluctuate without distinctive *trends* (in column 1) in a similar pattern. However, there is a significant at least order-1 time series autocorrelation (in columns 2 and 3) for all four time series.

9.4.2 Test for Non-Stationarity

To test whether there is a stationarity for these series, we can use the *Augmented Dickey Fuller Unit Root Tests for Stationary* which is implemented in *R* function *ur.df* as follows:

```
# Load the library urca
library(urca)
# Test for "OPEN"
test.OPEN = ur.df(dVIX$OPEN, type = "none",
                  selectlags = "AIC" )
# Print the summary of the test for "OPEN"
summary(test.OPEN)
```

```
##
## ###############################################
## # Augmented Dickey-Fuller Test Unit Root Test #
## ###############################################
##
## Test regression none
##
##
## Call:
## lm(formula = z.diff ~ z.lag.1 - 1 + z.diff.lag)
##
## Residuals:
##     Min      1Q  Median      3Q     Max
## -16.832  -0.672  -0.016   0.670  22.866
##
## Coefficients:
##             Estimate Std. Error t value Pr(>|t|)
## z.lag.1    -0.002574   0.000838   -3.07   0.0021 **
## z.diff.lag -0.158989   0.010748  -14.79   <2e-16 ***
## ---
## Signif. codes: 0 '***' 0.001 '**' 0.01 '*' 0.05 '.' 0.1 ' ' 1
##
## Residual standard error: 1.64 on 8437 degrees of freedom
## Multiple R-squared:  0.0268, Adjusted R-squared:  0.0265
## F-statistic:  116 on 2 and 8437 DF,  p-value: <2e-16
##
##
## Value of test-statistic is: -3.072
##
## Critical values for test statistics:
##       1pct  5pct 10pct
```

```
## tau1 -2.58 -1.95 -1.62

# Test for "HIGH"
test.HIGH = ur.df(dVIX$HIGH, type = "none",
                  selectlags = "AIC" )
# Print the summary of the test for "HIGH"
summary(test.HIGH)

##
## ###############################################
## # Augmented Dickey-Fuller Test Unit Root Test #
## ###############################################
##
## Test regression none
##
##
## Call:
## lm(formula = z.diff ~ z.lag.1 - 1 + z.diff.lag)
##
## Residuals:
##     Min      1Q  Median      3Q     Max
## -18.364  -0.654   0.003   0.662  25.110
##
## Coefficients:
##             Estimate Std. Error t value Pr(>|t|)
## z.lag.1    -0.003033   0.000851   -3.56  0.00037 ***
## z.diff.lag -0.012496   0.010886   -1.15  0.25103
## ---
## Signif. codes: 0 '***' 0.001 '**' 0.01 '*' 0.05 '.' 0.1 ' ' 1
##
## Residual standard error: 1.74 on 8437 degrees of freedom
## Multiple R-squared:  0.0017, Adjusted R-squared:  0.00146
## F-statistic: 7.18 on 2 and 8437 DF,  p-value: 0.000767
##
##
## Value of test-statistic is: -3.564
##
## Critical values for test statistics:
##       1pct  5pct 10pct
## tau1 -2.58 -1.95 -1.62

# Test for "LOW"
test.LOW = ur.df(dVIX$LOW, type = "none",
                 selectlags = "AIC" )
```

```
# Print the summary of the test for "LOW"
summary(test.LOW)

##
## ###############################################
## # Augmented Dickey-Fuller Test Unit Root Test #
## ###############################################
##
## Test regression none
##
##
## Call:
## lm(formula = z.diff ~ z.lag.1 - 1 + z.diff.lag)
##
## Residuals:
##     Min      1Q  Median      3Q     Max
## -23.925  -0.558  -0.019   0.561  20.545
##
## Coefficients:
##             Estimate Std. Error t value Pr(>|t|)
## z.lag.1    -0.002133   0.000726   -2.94   0.0033 **
## z.diff.lag -0.054599   0.010870   -5.02  5.2e-07 ***
## ---
## Signif. codes: 0 '***' 0.001 '**' 0.01 '*' 0.05 '.' 0.1 ' ' 1
##
## Residual standard error: 1.36 on 8437 degrees of freedom
## Multiple R-squared:  0.00412,    Adjusted R-squared:  0.00388
## F-statistic: 17.4 on 2 and 8437 DF,  p-value: 2.78e-08
##
##
## Value of test-statistic is: -2.936
##
## Critical values for test statistics:
##       1pct  5pct 10pct
## tau1 -2.58 -1.95 -1.62

# Test for "CLOSE"
test.CLOSE = ur.df(dVIX$CLOSE, type = "none",
                   selectlags = "AIC" )
# Print the summary of the test for "CLOSE"
summary(test.CLOSE)

##
## ###############################################
```

```
## # Augmented Dickey-Fuller Test Unit Root Test #
## ################################################
##
## Test regression none
##
##
## Call:
## lm(formula = z.diff ~ z.lag.1 - 1 + z.diff.lag)
##
## Residuals:
##     Min      1Q  Median      3Q     Max
## -16.818  -0.631  -0.029   0.618  22.706
##
## Coefficients:
##             Estimate Std. Error t value Pr(>|t|)
## z.lag.1    -0.002647   0.000839   -3.15   0.0016 **
## z.diff.lag -0.130768   0.010793  -12.12   <2e-16 ***
## ---
## Signif. codes: 0 '***' 0.001 '**' 0.01 '*' 0.05 '.' 0.1 ' ' 1
##
## Residual standard error: 1.63 on 8437 degrees of freedom
## Multiple R-squared:  0.0186, Adjusted R-squared:  0.0184
## F-statistic:   80 on 2 and 8437 DF,  p-value: <2e-16
##
##
## Value of test-statistic is: -3.154
##
## Critical values for test statistics:
##       1pct  5pct 10pct
## tau1 -2.58 -1.95 -1.62
```

As seen from the summary of the tests for these four time series, the values of test-statistic (in absolute values) are 3.0724 for *OPEN*, 3.5636 for *HIGH*, 2.9359 for *LOW*, and 3.1544 for *CLOSE*, respectively. All of them are lower than the three critical values for the test statistics (in absolute term) of 2.58(1pct), 1.95(5pct) and 1.62(10pct). So all these four time series are not stationary based on the *Dickey-Fuller test* and they are *I(1)* (i.e., integrated of order 1).

Although they are not a stationary series, they are in fact highly correlated. We can further explore the correlations among the four time series. For this purpose, we can make use of the *R* package *ggplot2* and its extension of *GGally* for *pairwise* plotting. As explained by its description, *the GGally extends ggplot2 by adding several functions to reduce the complexity of combining geometric objects with transformed data. Some of these functions include a pairwise plot matrix, a two-group pairwise plot matrix, a parallel coordinates plot, a survival plot, and several functions to plot networks*. This can be done by using the *ggpairs* function as follows:

```
# Load the "ggplot2"
library("ggplot2")
# Load the "GGally"
library("GGally")
```

```
## Registered S3 method overwritten by 'GGally':
##   method from
##   +.gg   ggplot2
```

```
# Call "ggpairs* from "GGally*
ggpairs(dVIX[,2:5])
```

```
## plot: [1,1] [=>-----------------------------------]  6%  est:0s
## plot: [1,2] [====>--------------------------------] 12% est:0s
## plot: [1,3] [======>------------------------------] 19% est:0s
## plot: [1,4] [=========>---------------------------] 25% est:0s
## plot: [2,1] [===========>-------------------------] 31% est:0s
## plot: [2,2] [=============>-----------------------] 38% est:0s
## plot: [2,3] [================>--------------------] 44% est:0s
## plot: [2,4] [==================>------------------] 50% est:0s
## plot: [3,1] [====================>----------------] 56% est:0s
## plot: [3,2] [=======================>-------------] 62% est:0s
## plot: [3,3] [=========================>-----------] 69% est:0s
## plot: [3,4] [===========================>---------] 75% est:0s
## plot: [4,1] [==============================>------] 81% est:0s
## plot: [4,2] [================================>----] 88% est:0s
## plot: [4,3] [==================================>--] 94% est:0s
## plot: [4,4] [=====================================]100% est:0s
##
```

As seen in Figure 9.5, these 4 time series are roughly correlated with statistically significant correlation coefficients more than 98%. This would lead to investigation of *cointegration analysis*

9.4.3 Vector Error Correction Model

We skip the *cointegration regression* since we would have 4 cointegration regressions depending on which of these 4 time series to be selected as the *response* variable to regress with the rest of the 3 time series. There is no financial theory to be based on on this selection and it can be arbitrary.

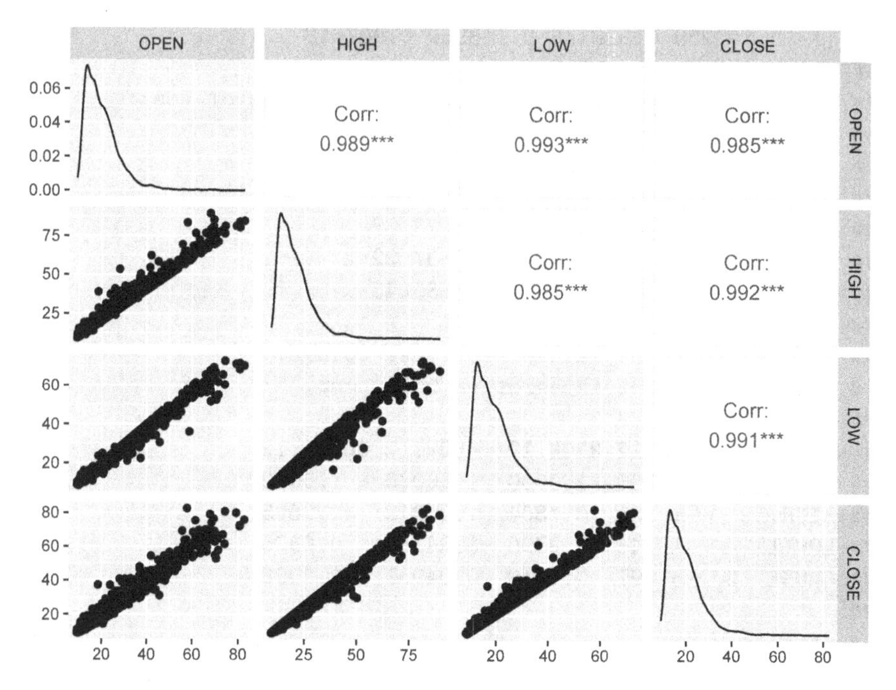

FIGURE 9.5
Pairwise Correlations among the Four Time Series

In the case of more than 2 financial time series to identify cointegration, the *Johansen Procedure for VAR* is the optimal testing procedure. This is implemented in the *R* function *ca.jo*. We can then use this *ca.jo* test procedure for the *VIX* data as follows:

```
# Load the "urca" library
library(urca)
# Perform ca.jo test
cajo.VIX = ca.jo(dVIX[,2:5], type="trace", spec="longrun")
# Print the summary of the test
summary(cajo.VIX)
```

```
##
## ######################
## # Johansen-Procedure #
## ######################
##
## Test type: trace statistic , with linear trend
##
## Eigenvalues (lambda):
```

```
## [1] 0.440239 0.333115 0.118959 0.007613
##
## Values of teststatistic and critical values of test:
##
##              test 10pct  5pct  1pct
## r <= 3 |   64.49  6.50  8.18 11.65
## r <= 2 | 1133.30 15.66 17.95 23.52
## r <= 1 | 4552.26 28.71 31.52 37.22
## r = 0  | 9448.96 45.23 48.28 55.43
##
## Eigenvectors, normalized to first column:
## (These are the cointegration relations)
##
##          OPEN.12 HIGH.12   LOW.12 CLOSE.12
## OPEN.12    1.000  1.0000  1.00000   1.0000
## HIGH.12    1.131 -0.7804  1.90918  -0.9595
## LOW.12     1.146 -0.8790 -3.24368   0.6326
## CLOSE.12  -3.297  0.6510  0.04957   1.1023
##
## Weights W:
## (This is the loading matrix)
##
##          OPEN.12  HIGH.12     LOW.12  CLOSE.12
## OPEN.d  -0.40766 -0.54576 -0.0008899 -0.003653
## HIGH.d  -0.27610  0.35805 -0.0648624 -0.008588
## LOW.d   -0.25595  0.16932  0.0559087 -0.006154
## CLOSE.d  0.04389  0.01557 -0.0022290 -0.010658
```

As seen from the output, the eigenvalues (*lambda*) generated by the test are (0.440239230, 0.333115391, 0.118959162, 0.007613084) with the largest eigenvalue of 0.440239230.

The next section shows the *Values of test statistic and critical values of test* for four hypotheses. For each of these four hypothesis tests, not only the statistic itself (i.e., column *test*) is given, but also the *critical values* at three levels of confidence of 10%, 5%, and 1% are given as shown below:

```
             test 10pct  5pct  1pct
r <= 3 |   64.49  6.50  8.18 11.65
r <= 2 | 1133.30 15.66 17.95 23.52
r <= 1 | 4552.26 28.71 31.52 37.22
r = 0  | 9448.96 45.23 48.28 55.43
```

The first hypothesis, $r = 0$, is to test for the presence of cointegration. It can be seen that there is a strong evidence to reject the null hypothesis of no cointegration since the value of the test statistic exceeds the 1% level significantly (i.e., *9448.96 > 55.43*).

The second test for null hypothesis of $r \leq 1$ against the alternative hypothesis of $r > 1$ also demonstrates clear evidence to reject the null hypothesis since the test statistic exceeds the 1% level significantly (i.e., *4552.26* $>$ *37.22*). This is also true in the third hypothesis test between the null hypothesis of $r \leq 2$ versus the alternative hypothesis of $r > 2$ since *1133.30* $>$ *23.52*.

The final hypothesis test for $r \leq 3$ against $r > 3$ also provides sufficient evidence for rejecting the null hypothesis since *64.49* $>$ *11.65*. This four tests conclude that the rank r of the matrix is greater than 3.

Therefore, the best estimate of the rank of the matrix would be 4, which means that we need a linear combination of these four *VIX* time series to form a stationary series. To form the linear combination, we take the information from the section on *Eigenvetors*:

```
Eigenvectors, normalized to first column:
(These are the cointegration relations)

            OPEN.l2    HIGH.l2       LOW.l2   CLOSE.l2
OPEN.l2    1.000000  1.0000000  1.00000000  1.0000000
HIGH.l2    1.131226 -0.7803826  1.90918085 -0.9595074
LOW.l2     1.145636 -0.8789760 -3.24367879  0.6325538
CLOSE.l2 -3.296502  0.6510442  0.04956708  1.1022933
```

We make use of the *eigenvector* components associated with the largest eigenvalue of 0.440239230 in the first column named *OPEN.l2* to form a new time series (let's name is as *nVIX*) as follows:

```
# Form the new linear combination series
nVIX = dVIX$OPEN + 1.131226*dVIX$HIGH +
  1.145636*dVIX$LOW -3.296502*dVIX$CLOSE
```

This new combined time series *nVIX* should be a stationary series. We can use *Augmented Dickey-Fuller Test Unit Root Test* to confirm this finding as follows:

```
# Call the Dickey-Fuller Test
test.nVIX = ur.df(nVIX, type = "none",
                  selectlags = "AIC" )
# Print the test summary
summary(test.nVIX)
```

```
##
## ###############################################
## # Augmented Dickey-Fuller Test Unit Root Test #
## ###############################################
```

```
##
## Test regression none
##
##
## Call:
## lm(formula = z.diff ~ z.lag.1 - 1 + z.diff.lag)
##
## Residuals:
##     Min      1Q Median      3Q     Max
## -49.82  -1.04  -0.04   1.20  34.37
##
## Coefficients:
##            Estimate Std. Error t value Pr(>|t|)
## z.lag.1     -1.1842     0.0165  -71.63   <2e-16 ***
## z.diff.lag   0.0263     0.0109    2.41    0.016 *
## ---
## Signif. codes: 0 '***' 0.001 '**' 0.01 '*' 0.05 '.' 0.1 ' ' 1
##
## Residual standard error: 3.01 on 8437 degrees of freedom
## Multiple R-squared:  0.577,  Adjusted R-squared:  0.577
## F-statistic: 5.76e+03 on 2 and 8437 DF,  p-value: <2e-16
##
##
## Value of test-statistic is: -71.63
##
## Critical values for test statistics:
##       1pct  5pct 10pct
## tau1 -2.58 -1.95 -1.62
```

It can be seen that the test statistic value of 71.6274 (in absolute value) is higher than of 3 critical values of 2.58 (1pct), 1.95(5pct), and 1.62(10pct), indicating stationarity of $nVIX$. This should be true since these four time series are created from the daily S&P 500 Index (SPX) using *OPEN*, *HIGH*, *LOW*, and *CLOSE* values and altogether coincide and track the S&P 500 index. Therefore, they should be correlated and form a long-term relationship among these four time series.

Through stationary, the new series $nVIX$ still exhibits some amount of autocorrelation and may have a long-term memory, which can be graphically demonstrated in Figure 9.6 produced with the following *R* code chunk:

```
# Figure layout
layout(matrix(c(1,1,2,3),2,2,byrow=T))
# Add nVIX to the dataframe dVIX
dVIX$nVIX = nVIX
# Plots for "nVIX"
```

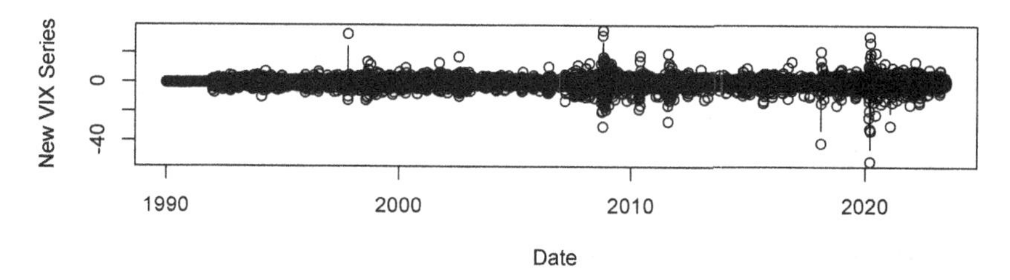

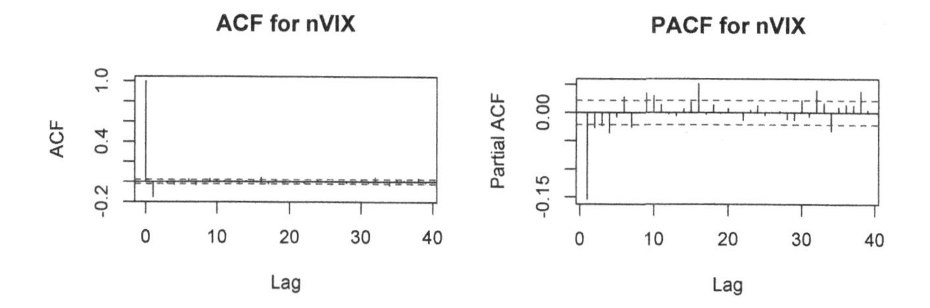

FIGURE 9.6
Time Series Plot for the New VIX Series from Johansen Procedure for VAR

```
plot(nVIX~Date, dVIX,type="b",
     xlab="Date",ylab="New VIX Series",
     main="Time Series Plot for the New Johansen Procedure Series")
acf(nVIX, main="ACF for nVIX");
pacf(nVIX, main="PACF for nVIX")
```

9.5 Discussions

This chapter served as a preliminary transition from *ARIMA* and *GARCH* modeling to *cointegration analysis. Cointegration analysis* extends the classical *ARIMA* and *GARCH* analyses to multiple time series and is especially relevant when exploring the long-term relationships between them. We hope that interested readers have learned that it can be particularly useful when dealing with assets or variables that are expected to have a stable, long-term relationship, despite exhibiting short-term deviations or divergences.

9.5.1 How to Do Cointegration in *R*

Cointegration analysis in *R* can be performed using different *R* packages, with the most common ones being **stats**, **urca**, and **vars**. The general recommendation on how to conduct *cointegration analysis* using these packages can be outlined below.

1. *Install* and *Load* the Required Packages:

   Before cointegration analysis, make sure that the necessary *R* packages are installed. If not, use the *install.packages()* function to install them. We use *urca* package as example, but interested readers can select any *R* packages you are most familiar with:

   ```
   # To install "urca" package
   install.packages("urca")
   ```

   Then, load the installed *R* package:

   ```
   # Load the "urca"
   library(urca)
   ```

2. Prepare the Data:

   As shown in the example above, read in the data and create time series data frame or a time series object. Make sure that the data frame has multiple variables (time series) to be investigated for cointegration.

3. Conduct *Cointegration* Tests:

   Two *cointegration tests* in *R* can be used with the first test as the *Engle-Granger test* and the second test as the *Johansen test.*

   The *Engle-Granger test* can be used for cointegration between two series. This can be performed using the *ca.jo* function with options *type = "eigen"* and *K = 2.*

   The *Johansen test* is used to test cointegration among more than two-time series. The typical *R* code chunk would be as follows:

   ```
   # Call ca.jo function for the data "df"
   # with multiple variables
   cajo.df = ca.jo(df, type = "trace", K = 2)
   # Print the summary of the test
   summary(cajo.df)
   ```

In the above code chunk, *ca.jo()* performs the *Johansen test. type = "trace"* specifies the method (trace statistics), and K is the maximum number of cointegrating relationships to be tested. *summary(cajo.df)* displays the test results, including critical values and cointegration ranks.

Both the *Engle-Granger* and *Johansen tests* provide outputs with test statistics, critical values, and cointegration ranks, which are to be used for determining whether cointegration exists and how many cointegrating relationships exist.

4. Model the Cointegrated System:

   If cointegration is confirmed, we can proceed to model the cointegrated system using an *Error Correction Model (ECM)* or a *Vector Error Correction Model (VECM)* for further financial analysis.

9.5.2 Cointegration in Financial Analysis

Cointegration is a crucial concept in financial time series analysis to understand the long-term relationships between multiple financial assets or economic variables. It was first introduced by *Clive Granger*, a Nobel laureate in economics, and *Robert Engle* in the early 1980s.

Financial markets are well known for their complex and interrelated nature. Many financial time series data, such as stock prices or exchange rates, exhibit trends and fluctuations over time. *Cointegration* provides a framework for analyzing such data and identifies whether multiple time series move together in the long run. In essence, *cointegration* can provide insights that, while individual financial time series might not directly be correlated or stationary, a well-organized linear combination of these series can be stationary. Therefore, *cointegrated time series* can have a stable, long-term relationship, even if they may display short-term volatility or divergence.

The key components of cointegration in financial time series analysis can include:

1. Stationarity:

   Stationarity is the key concept to be achieved for *cointegration*. A stationary time series is one whose statistical properties, such as mean and variance, remain constant over time. Financial time series data are often *non-stationary* and exhibit *trends, seasonality*, or other patterns that change over time.

2. Cointegration Vector:

   Cointegration vector is then used to construct a linear combination of non-stationary time series, so to achieve stationarity. It represents the equilibrium or long-term relationship between the variables. For example, if there are three stocks, A, B, and C, and they are cointegrated, there exists a linear combination, say $nABC = A - 2B + 3C$, that is stationary, where the coefficients of (1,-2,3) are the *cointegration vector* from *ECM* model.

3. Error Correction Model (ECM):

   The *Error Correction Model (ECM)* is one of the practical models of cointegration in financial analysis. The *ECM* is a regression model that incorporates both the short-term dynamics and the long-term equilibrium represented by the *cointegration vector*. It helps to explain how the variables adjust when they deviate from their long-term relationship.

4. Trading and Investment Strategies:

   Cointegration has significant implications for traders and investors in financial market. If two or more financial assets are cointegrated, deviations from their long-term relationship can provide trading opportunities. When the spread between the assets widens or narrows significantly, traders may consider taking positions to profit from mean reversion, expecting the spread to return to its historical relationship.

5. Risk Management:

   Understanding *cointegration* can also be valuable for risk management. Investors can diversify their portfolios by selecting assets that are not only individually promising but also cointegrated with each other. This diversification strategy can potentially reduce risk.

In conclusion, *cointegration* is a vital tool in financial time series analysis that helps analysts identify and exploit relationships among assets or variables that persist in the long term, even when short-term fluctuations may exist. It plays a critical role in statistical arbitrage, risk management, and portfolio optimization in the world of finance. Analysts use various statistical tests, such as the *Engle-Granger test* or *Johansen test*, to determine whether cointegration exists between financial time series. It's important to note that cointegration does not imply causality, but rather a long-term relationship that can be exploited for various financial purposes.

9.5.3 Cointegration in Investment among Stocks to Mutual Funds

As seen in this chapter, *cointegration* is a statistical concept used to analyze the long-term relationships between two or more financial time series. It has been employed in econometrics and finance to assess whether two or more assets or financial instruments move together in the long run. Although *cointegration* itself does not provide a theoretical basis for selecting and switching from stocks to mutual funds or vice versa, it instead helps investors understand the long-term relationships and dependencies between different assets.

Here's how *cointegration* might be relevant and helpful in the context of stocks and mutual funds:

1. *Diversification*: One of the primary reasons investors choose to invest in mutual funds is to achieve diversification by holding a portfolio of different stocks or other securities. *Cointegration analysis* can be used to identify assets that tend to move together over time, helping investors select mutual funds that provide effective diversification.
2. *Risk Management*: *Cointegration* can help investors manage risk by identifying financial assets that are likely to be positively or negatively correlated over time. For investors with a portfolio of stocks who want to reduce risk, they might consider adding mutual funds with *cointegrated* assets to offset stock price fluctuations.
3. *Portfolio Optimization*: For best investment return, *cointegration analysis* can be part of a broader portfolio optimization strategy. It can help investors and wealth managers build portfolios that balance risk and return based on historical relationships between assets.
4. *Market Timing*: Interested investors can use *cointegration analysis* to make market timing decisions. As an example, if two assets have cointegrated behavior and have temporarily diverged, one may switch from one to the other with the expectation that these two assets will revert to their long-term relationship.

It's important in financial market and analysis on investment among financial stocks and mutual funds depending on individual financial goals, risk tolerance, and investment strategy. *Cointegration analysis* can provide insights into the relationships between assets as one of the investment decision-making toolkits.

10

Financial Statistical Modeling in Risk and Wealth Management

Financial statistical modeling is a cornerstone of both risk management and wealth management. It equips financial professionals in these fields with the tools to make data-driven decisions and risk assent, as well as evidence-informed and optimized financial strategies. With all the financial modeling and statistical analysis from Chapters 1 to 9, this chapter concludes this book to give a brief discussion and introduction of the critical role of statistical financial modeling in risk and wealth management.

10.1 Statistical Modeling in Risk Management

Risk management is a critical aspect of financial decision-making. Financial institutions, investors, and businesses need to assess, measure, and mitigate risks to ensure the stability and profitability of their operations. To achieve this, financial statistics play a vital role in quantifying and understanding the various risks associated with financial assets and portfolios.

10.1.1 Understanding Risk

Risk in finance refers to the uncertainty or variability of returns on an investment. It encompasses a wide range of factors, including market fluctuations, credit defaults, interest rate changes, and more. Financial statistics and financial data analysis are used to assess and manage these risks effectively. There are several types of financial risk to be considered:

1. *Market Risk*: This is the risk that arises from market price fluctuations, including stock prices, interest rates, and exchange rates. Key statistics for measuring market risk include volatility, beta, and value at risk (*VaR*).

DOI: 10.1201/9781003469704-10

2. *Credit Risk*: Credit risk relates to the potential for borrowers to default on their obligations. Key statistics include credit ratings, default probability, and loss given default (LGD).
3. *Interest Rate Risk*: Interest rate risk pertains to changes in interest rates affecting the value of fixed-income securities. Yield duration and convexity are crucial statistics for measuring this risk.
4. *Operational Risk*: Operational risk encompasses the risk of losses resulting from inadequate or failed internal processes, systems, or human errors. Statistical tools like loss distribution and event frequency analysis are used to assess this risk.
5. *Liquidity Risk*: Liquidity risk is the risk of being unable to buy or sell assets quickly without significant price changes. Liquidity ratios, trading volumes, and bid-ask spreads are relevant statistics.

10.1.2 Key Financial Statistics

Key financial statistics associated with financial risk are critical for assessing, managing, and mitigating risks in various financial domains. These statistics provide insights into the potential volatility and downside exposure of financial instruments, portfolios, and organizations. Here are some of the key financial statistics commonly used in financial risk analysis:

1. *Standard Deviation*: Standard deviation measures the dispersion or volatility of returns in a financial dataset. Higher standard deviation indicates greater variability, which is a key component of risk assessment.
2. *Volatility*: Volatility is another statistical measure of the dispersion of returns for a financial asset. It is often represented by the standard deviation of an asset's returns. A higher standard deviation indicates greater price variability, implying higher risk. Investors often use volatility to assess the potential downside of an investment. Managing volatility is critical in portfolio diversification and asset allocation strategies.
3. *Correlation*: Correlation measures the degree to which two financial assets move in relation to each other. Negative correlation can help reduce portfolio risk through diversification.
4. *Covariance*: Covariance quantifies how two assets move together. Positive covariance indicates that the assets tend to move in the same direction, while negative covariance implies they move in opposite directions.

5. *Beta*: Beta is a measure of an asset's sensitivity to market movements. It quantifies the asset's risk relative to the overall market. A beta of 1 indicates the asset moves in line with the market, while a beta greater than 1 implies higher volatility, and a beta less than 1 suggests lower volatility. Beta is an essential tool in the construction of diversified portfolios and the assessment of systematic risk.

6. *Value at Risk (VaR)*: VaR is a fundamental statistic that quantifies the potential loss a portfolio or financial instrument may incur over a specified time horizon at a given confidence level. It is a statistical tool used to estimate the maximum potential loss an investment or portfolio could incur over a specified time horizon at a given confidence level. It provides a numerical estimate of downside risk. Common confidence levels include 95% and 99%. VaR is crucial in setting risk limits, determining capital adequacy, and managing overall risk exposure.

7. *Tracking Error*: Tracking error assesses the deviation of a portfolio's returns from its benchmark index. Higher tracking error suggests increased risk and potential for underperformance relative to the benchmark.

8. *Drawdown*: Drawdown is the peak-to-trough decline in the value of a portfolio or asset. Understanding the magnitude and duration of drawdowns is essential for risk management.

9. *Liquidity Metrics*: Metrics such as bid-ask spreads, trading volumes, and market depth provide insights into an asset's liquidity risk. Assets with wider spreads and lower trading volumes are often riskier.

10. *Duration and Convexity*: For managing interest rate risk, duration measures the sensitivity of a fixed-income investment, such as bond's price, to changes in interest rates. It helps assess the risk associated with bond investments. Longer durations imply higher interest rate risk, while shorter durations indicate lower risk. Convexity is a measure of the curvature of the bond's price-yield curve, indicating how price changes with varying interest rates. Understanding duration and convexity helps in making prudent fixed-income investment decisions and managing interest rate exposure.

11. *Loss Given Default (LGD)*: LGD represents the expected loss in case of a credit default. It is expressed as a percentage of the principal amount and is a crucial metric in credit risk analysis. LGD is vital in the determination of capital requirements, setting credit limits, and assessing the overall credit portfolio's health.

12. *Solvency Ratios*: Solvency ratios, like the debt-to-equity ratio or interest coverage ratio, assess the financial health and ability of an organization to meet its financial obligations.

13. *Sharpe Ratio*: The Sharpe ratio evaluates the risk-adjusted performance of an investment or portfolio. It considers both returns and volatility, helping investors assess whether the risk taken is adequately compensated by returns.

14. *Sortino Ratio*: Similar to the Sharpe ratio, the Sortino ratio focuses on downside risk, considering only the volatility of negative returns. It provides a more refined risk assessment, particularly for risk-averse investors.

15. *Earnings Before Interest and Taxes (EBIT) Margin*: EBIT margin reflects a company's profitability, helping assess its capacity to withstand financial downturns.

16. *Credit Metrics*: Credit metrics like credit ratings and credit spreads are crucial for assessing the credit risk associated with bonds and fixed-income instruments.

17. *Gearing Ratio*: The gearing ratio measures the financial leverage of a company. High gearing implies greater financial risk.

18. *Market-to-Book Ratio*: This ratio assesses the market value of a company relative to its book value. A high market-to-book ratio may indicate a risk of overvaluation.

19. *Financial Stress Indicators*: Metrics like the VIX (Volatility Index) or credit default swap (CDS) spreads provide insight into market stress and credit risk.

20. *Regulatory Capital Ratios*: Regulatory capital ratios, such as the Basel III capital adequacy ratios for banks, help ensure they maintain adequate capital buffers to absorb potential losses.

21. *Credit Ratings*: Credit ratings are assessments of the creditworthiness of borrowers or debt issuers. Rating agencies assign letter grades such as AAA, AA, A, and so on to indicate the level of credit risk associated with a particular entity or security. Understanding credit ratings is essential for making informed investment decisions, especially in the fixed-income market and when dealing with debt securities.

These key financial statistics provide a comprehensive framework for evaluating and managing financial risk across different sectors of the financial industry, including investment management, banking, insurance, and corporate finance.

Proper risk assessment and management often involve a combination of these metrics to gain a holistic view of financial risk.

10.1.3 Statistical Tools and Models

Risk management employs various statistical tools and models to assess, model, and mitigate risks effectively. Besides the commonly-used statistical models discussed in Chapters 2 to Chapter 9, other financial-specific models include the Black-Scholes model for option pricing, the Capital Asset Pricing Model (CAPM) for assessing market risk, and credit risk models like the CreditMetrics and Moody's KMV models for credit risk assessment. Specifically,

1. *Black-Scholes Model*: The Black-Scholes model is a widely used formula for pricing options. It takes into account factors such as the underlying asset's price, the option's strike price, the time to expiration, the risk-free interest rate, and the asset's volatility. Understanding the Black-Scholes model is crucial for valuing options and managing option-based risks in investment portfolios.
2. *Capital Asset Pricing Model (CAPM)*: CAPM is a model that describes the relationship between systematic risk and expected return for assets, particularly stocks. It helps investors determine whether an investment provides adequate returns relative to its risk. CAPM is valuable in setting required rates of return for investments and evaluating the performance of portfolios in relation to market risk.
3. *CreditMetrics*: CreditMetrics is a statistical tool developed by J. P. Morgan that measures credit risk within a portfolio. It uses statistical techniques to estimate the potential credit losses and the credit value at risk. CreditMetrics aids in understanding the potential downside associated with credit exposures and assists in setting appropriate credit risk management strategies.
4. *Moody's KMV Model*: Moody's KMV model is a sophisticated tool for estimating the probability of default for a company based on its market value of assets and liabilities. It provides insights into the creditworthiness of a firm and aids in making informed credit decisions. Understanding the Moody's KMV model is essential for credit risk analysts and professionals working in financial institutions.

In summary, financial statistics are indispensable for risk management in the world of finance. By understanding and effectively utilizing statistical measures and models, financial professionals can make informed decisions, reduce exposure to risk, and enhance the overall performance and stability of their investments and portfolios. This section has provided an introductory overview of some key statistical concepts and tools used in the field of financial risk management. However, it is important to delve deeper into each of these areas for a more comprehensive understanding of financial statistics and their application in risk management. Successful risk management requires a robust understanding of statistical principles and their application in real-world financial scenarios.

For more discussions and learning on financial modeling in risk management, we refer to the great book by McNeil et al. (2015), which provides the most comprehensive treatment from the theoretical concepts to quantitative modeling techniques in real-life risk management.

10.2 Statistical Modeling in Wealth Management

Wealth management is a multifaceted field, and it relies heavily on financial statistical modeling to make informed decisions, mitigate risks, and achieve financial objectives. Successful wealth management involves the careful stewardship of assets and financial resources to achieve long-term financial goals. To make informed decisions and provide personalized financial solutions, wealth managers rely on a variety of financial statistical models and techniques. We explore here the significance of financial statistical modeling in wealth management and provides insights into its practical applications.

10.2.1 The Role of Financial Statistical Modeling

Financial statistical modeling in wealth management serves several critical purposes:

1. *Risk Assessment*: It helps assess and quantify risks associated with investment strategies and portfolios. By analyzing historical data and utilizing statistical models, wealth managers can estimate the potential downside risks, such as market volatility and drawdowns.
2. *Asset Allocation*: Statistical models assist in optimizing asset allocation decisions. These models consider factors like expected returns, risk, and correlations among asset classes to design portfolios that align with clients' financial objectives and risk tolerance.
3. *Return Forecasting*: Wealth managers use statistical models to project future returns for various asset classes. Projections are based on historical data, economic indicators, and other relevant factors. These forecasts guide investment decisions.
4. *Portfolio Optimization*: Modern Portfolio Theory (MPT) employs mathematical and statistical tools to construct portfolios that aim

to maximize returns while minimizing risk. Efficient frontiers, mean-variance optimization, and other techniques are used to build diversified and efficient portfolios.

5. *Client Profiling*: Wealth managers use statistical models to create financial profiles of their clients. These profiles help in understanding individual risk tolerance, investment horizon, and financial goals, enabling personalized wealth management strategies.

10.2.2 Key Statistical Models in Wealth Management

There are numerous statistical models used in wealth management. To list a few here:

1. *Modern Portfolio Theory (MPT)*: MPT, developed by Harry Markowitz, is a fundamental model in wealth management. It uses statistical tools to construct portfolios that offer the best trade-off between risk and return. By diversifying across multiple assets, MPT aims to achieve efficient portfolios that maximize returns for a given level of risk.

2. *Capital Market Expectations*: Wealth managers employ statistical models to develop capital market expectations, which involve estimating the future performance of various asset classes. These expectations are derived from historical data, macroeconomic factors, and financial market indicators.

3. *Monte-Carlo Simulation*: Monte-Carlo simulation is widely used for scenario analysis. It involves generating multiple random scenarios to assess how different variables, such as investment returns and inflation rates, can impact a portfolio's performance over time. This statistical approach helps wealth managers develop robust financial plans and strategies.

4. *Value at Risk (VaR)*: VaR, as discussed in previous chapters, is used to estimate potential losses within a portfolio at a given confidence level. It aids wealth managers in setting risk limits and developing strategies to protect clients' assets from extreme market events.

5. *Regression Analysis*: Regression analysis is employed to analyze the relationship between different asset classes or investment strategies. Wealth managers use this statistical method to assess the impact of variables like interest rates, economic indicators, or market conditions on the performance of investments.

10.2.3 Portfolio Construction and Optimization

Statistical financial modeling plays a pivotal role in constructing and optimizing investment portfolios that align with clients' financial objectives:

1. *Modern Portfolio Theory (MPT)*: Wealth managers apply statistical modeling to construct diversified portfolios that aim to maximize returns for a given level of risk. These models identify optimal asset allocations, including stocks, bonds, and alternative investments, to meet clients' financial goals. For example, a wealth manager designs a diversified portfolio for a client seeking growth with moderate risk tolerance. The model uses historical return data to determine an optimal allocation between equities, fixed income, and alternative investments that aligns with the client's risk profile.
2. *Capital Market Expectations*: Statistical models are employed to estimate the future performance of asset classes. These models consider historical data, economic indicators, and market trends to inform portfolio decisions. Capital market expectations can guide the decision to overweight or underweight specific asset classes in a client's portfolio. As an example, a wealth manager incorporates economic indicators and market data into a statistical model to estimate the future performance of asset classes. These expectations indicate that emerging markets are expected to outperform developed markets over the next five years, leading to an overweight allocation in emerging market equities within a client's portfolio.
3. *Monte-Carlo Simulation*: Monte-Carlo simulations use statistical modeling to assess the probability of achieving clients' financial goals under various scenarios. Factors such as investment returns, inflation, and life expectancy are considered in these simulations, making them valuable for financial planning and risk mitigation. For example, a wealth manager runs a Monte-Carlo simulation for a client planning for retirement. By considering different scenarios, including variations in investment returns and inflation rates, the simulation assesses the likelihood of meeting the client's retirement objectives. This analysis guides financial planning and risk management strategies.

In summary, financial statistical modeling is at the heart of wealth management, allowing professionals to make well-informed decisions, tailor investment strategies, and navigate the complexities of financial markets. These models play a pivotal role in risk assessment, asset allocation, return forecasting, and client profiling.

By understanding and effectively using financial statistical models, wealth managers can provide clients with the confidence that their financial objectives are being pursued while minimizing risk exposure. Staying current with evolving statistical methodologies is essential to continue delivering value in an ever-changing financial landscape.

For more discussion and learning on financial modeling in wealth management, we recommend the book by Day (2012). In this book, The author, *Alastair Day*, did a great job of making complex issues simple and easy to understand with implementation in *Excel.* This book covered the comprehensive tools and methods in financial modeling using Microsoft Excel to enable you for better wealth management. We also recommend the book by Rees (2008) for financial professionals in *intermediate and advanced level.* Based on the author's extensive experience in business and financial modeling, this book covered an extensive topics in financial modeling and wealth management.

10.3 Discussions and Recommendations

Statistical financial modeling is a versatile and indispensable tool in both risk management and wealth management. It enables professionals in these fields to make well-informed decisions, manage risks, and develop customized financial strategies. Understanding the dynamic nature of financial markets, staying current with evolving modeling methodologies, and continually enhancing modeling techniques are essential for success in both risk management and wealth management. It empowers wealth managers to make informed decisions, design personalized investment strategies, and navigate the complex landscape of financial markets. By leveraging a range of statistical models and techniques, wealth managers can provide clients with wealth preservation, growth, and financial security while aligning their portfolios with specific financial goals and risk tolerances. Understanding and using these models effectively is essential in an industry where precision and client-centered solutions are paramount.

The integration of financial statistical modeling into risk and wealth management ensures that clients' financial objectives are met while mitigating risks and capitalizing on opportunities. Risk and wealth managers must remain vigilant in staying abreast of evolving statistical methodologies to continue delivering value to their clients in a constantly changing financial landscape. This book hopes to contribute to the knowledge and learning in financial modeling to the risk and wealth managements.

Concluding Remarks:

More Topics on Financial Modeling to Risk and Wealth Management Will Be Discussed in the Next Book! SO Expect SOON!

THANK YOU FOR YOUR INTEREST IN THIS BOOK!!!

Bibliography

Adler, J. (2012). *R In a Nutshell*, 2nd Edition. O'Reilly Media, Inc., Sebastopol, CA.

Bates, D. M. and Watts, D. G. (1998). *Nonlinear Regression Analysis and Its Applications.* Wiley.

Bollerslev, T. (1986). Generalized autoregressive conditional heteroskedasticity. *Journal of Econometrics*, 31(3):307–327.

Chambers, J. M. (1998). *Programming with Data.* Springer, New York, USA.

Chambers, J. M. (2008). *Software for Data Analysis: Programming with R.* Springer, New York, USA.

Day, A. (2012). *Mastering Financial Modelling in Microsoft Excel: A Practitioner's Guide to Applied Corporate Finance (The Mastering Series)*, 3rd Edition. FT Publishing International.

DeGroot, M. H. and Schervish, M. J. (2011). *Probability and Statistics*, 4th Edition. Pearson.

Dobson, A. J. and Barnett, A. G. (2018). *An Introduction to Generalized Linear Models*, 3rd Edition. Chapman & Hall/CRC Texts in Statistical Science.

Dunn, P. K. and Smyth, G. K. (2018). *Generalized Linear Models with Examples in R.* Springer, New York.

Engle, R. F. (1982). Autoregressive conditional heteroscedasticity with estimates of the variance of United Kingdom inflation. *Econometrica*, 50(4): 987–1007.

Everitt, B. and Hothorn, T. (2006). *A Handbook of Statistical Analyses Using R.* Chapman & Hall/CRC, Boca Raton, FL.

Faraway, J. J. (2004). *Linear Models with R.* Chapman & Hall/CRC, Boca Raton, FL. ISBN 1-584-88425-8.

Faraway, J. J. (2006). *Extending Linear Models with R: Generalized Linear, Mixed Effects and Nonparametric Regression Models.* Chapman & Hall/CRC, Boca Raton, FL. ISBN 1-584-88424-X.

Francis, A. (2004). *Business Mathematics and Statistics*, 6th Edition. Cengage Learning, Hampshire, U.K.

Gardener, M. (2012). *Beginning R: The Statistical Programming Language.* John Wiley & Sons, Inc., Indianapolis, IN.

Ihaka, R. and Gentleman, R. (1996). R: A language for data analysis and graphics. *Journal of Computational and Graphical Statistics*, 5(3):299–314.

James, G., Witten, D., Hastie, T., and Tibshirani, R. (2023). *An Introduction to Statistical Learning with Applications in R*, 2nd Edition. Springer-Verlag, New York.

Kabacoff, R. I. (2011). *R In Action: Data Analysis and Graphics with R.* Manning Publications Co, New York.

Kleiber, C. and Zeileis, A. (2008). *Applied Econometrics with R.* Springer-Verlag, New York.

Krämer, W. and Sonnberger, H. (1986). *The Linear Regression Model under Test.* Physica, Heidelberg.

Ljung, G. M. and Box, G. E. P. (1978). On a measure of lack of fit in time series models. *Biometrika*, 65:297–303.

McCullagh, P. and Nelder, J. A. (1995). *Generalized Linear Models*, 2nd Edition. Chapman and Hall.

McNeil, A. J., Frey, R., and Embrechts, P. (2015). *Quantitative Risk Management: Concepts, Techniques and Tools*, Revised Edition. Princeton University Press.

Murrell, P. (2005). *R Graphics.* Chapman & Hall/CRC, Boca Raton, FL.

Nash, J. C. (2014). *Nonlinear Parameter Optimization Using R Tools.* Wiley, John Wiley & Sons Ltd, West Sussex, United Kingdom.

Nicols, D. A. (1983). Macroeconomic determinants of wage adjustments in white collar occupations. *Review of Economics and Statistics*, 65:203–213.

Pfaff, B. (2008). *Analysis of Integrated and Cointegrated Time Series with R*, 2nd Edition. Springer, New York. ISBN 0-387-27960-1.

Rees, M. (2008). *Financial Modelling in Practice: A Concise Guide for Intermediate and Advanced Level*, 1st Edition. Wiley.

Rizzo, M. L. (2008). *Statistical Computing with R.* Chapman & Hall/CRC, Boca Raton, FL.

Ruppert, D. and Matteson, D. S. (2015). *Statistics and Data Analysis for Financial Engineering with R Examples*, 2nd Edition. Springer.

Sarkar, D. (2008). *Lattice: Multivariate Data Visualization with R.* Springer, New York.

Tao, R. and Brooks, C. (2019). *Python Guide to Accompany Introductory Econometrics for Finance (October 25, 2019).* Available at SSRN: https://ssrn.com/abstract=3475303 or http://dx.doi.org/10.2139/ssrn.3475303.

Wackerly, D., Mendenhall, W., and Scheaffer, R. L. (2008). *Mathematical Statistics with Applications*, 7th Edition. Brooks/Cole, Cengage Learning.

Wasserman, L. (2003). *All of Statistics: A Concise Course in Statistical Inference.* Springer.

Wickham, H. (2016). *ggplot2: Elegant Graphics for Data Analysis (Use R)*, 2nd Edition. Springer.

Xie, Y. (2019). *bookdown: Authoring Books and Technical Documents with R Markdown.* R package version 0.13.

Index

S

T

U

V

W

Printed in the United States
by Baker & Taylor Publisher Services